Professor Stewart's
Hoard of
Mathematical Treasures

By the Same Author

Concepts of Modern Mathematics
Game, Set, and Math
Does God Play Dice?
Another Fine Math You've Got Me Into
Fearful Symmetry (with Martin Golubitsky)
Nature's Numbers
From Here to Infinity
The Magical Maze
Life's Other Secret
Flatterland
What Shape Is a Snowflake?
The Annotated Flatland (with Edwin A. Abbott)
Math Hysteria
The Mayor of Uglyville's Dilemma
Letters to a Young Mathematician
How to Cut a Cake
Why Beauty Is Truth
Taming the Infinite
Professor Stewart's Cabinet of Mathematical Curiosities

with Jack Cohen
The Collapse of Chaos
Figments of Reality
What Does a Martian Look Like?
Wheelers (science fiction)
Heaven (science fiction)

with Terry Pratchett and Jack Cohen
The Science of Discworld
The Science of Discworld II: The Globe
The Science of Discworld III: Darwin's Watch

Professor Stewart's Hoard of Mathematical Treasures

Ian Stewart

BASIC BOOKS

A Member of the Perseus Books Group
New York

Hardcover first published in Great Britain in 2009 by
PROFILE BOOKS LTD
3A Exmouth House
Pine Street
London WEC1R 0JH
www.profilebooks.com
Paperback first published in 2010 by Basic Books,
A Member of the Perseus Books Group

Books published by Basic Books are available at special discounts for
bulk purchases in the United States by corporations, institutions, and
other organizations. For more information, please contact the Special
Markets Department at the Perseus Books Group, 2300 Chestnut Street,
Suite 200, Philadelphia, PA 19103, or call (800) 810-4145, ext. 5000, or
e-mail special.markets@perseusbooks.com.

Cataloging-in-Publication data for this book is available from the
Library of Congress.

Hardcover ISBN: 978-1-84668-292-6
Paperback ISBN: 978-0-465-01775-1

10 9 8 7 6 5 4 3 2 1

Contents

Acknowledgements

The following figures are reproduced with the permission of the named copyright holders:

Pages 30, 280 ('What Seamus Didn't Know'); Suppiya Siranan.

Page 41 ('What is the Area of an Ostrich Egg?'); Hierakonpolis expedition, leader Renée Friedman, photograph by James Rossiter.

Page 69 ('Mathematical Cats'); Dr Sergey P. Kuznetsov, Laboratory of Theoretical Nonlinear Dynamics, SB IRE RAS.

Page 92 ('How to See Inside Things'); Brad Petersen.

Page 107 ('Alexander's Horned Sphere'); from *Topology* by John G. Hocking and Gail S. Young, Addison-Wesley, 1961.

Page 113 ('Just a Phase I'm Going Through'); GNU Free Documentation License, Free Software Foundation (www.gnu.org/copyleft/fdl.html).

Page 182 ('The Klein Bottle'); Janet Chao (www.illustrationideas.com).

Page 184 ('The Klein Bottle'); Konrad Polthier, Free University of Berlin.

Page 190 ('Multiplying with Sticks'); Eric Marcotte PhD (www.sliderule.ca).

Page 216 ('How to Turn a Sphere Inside Out'); Bruce Puckett.

Second Drawer Down ...

When I was fourteen, I started collecting mathematical curiosities. I've been doing that for nearly fifty years now, and the collection has outgrown the original notebook. So when my publisher suggested putting together a mathematical miscellany, there was no shortage of material. The result was *Professor Stewart's Cabinet of Mathematical Curiosities*.

Cabinet was published in 2008, and, as Christmas loomed, it began to defy the law of gravity. Or perhaps to obey the law of levity. Anyway, by Boxing Day it had risen to number 16 in a well-known national bestseller list, and by late January it had peaked at number 6. A mathematics book was sharing company with Stephenie Meyer, Barack Obama, Jamie Oliver and Paul McKenna.

This was, of course, completely impossible: everyone knows that there aren't that many people interested in mathematics. Either my relatives were buying a huge number of copies, or the conventional wisdom needed a rethink. So then I got an email from my publisher asking whether there might be any prospect of a sequel, and I thought, 'My suddenly famous Cabinet is still bursting at the seams with goodies, so why not?' *Professor Stewart's Hoard of Mathematical Treasures* duly emerged from darkened drawers into the bright light of day.

It's just what you need to while away the hours on your desert island. Like its predecessor, you can dip in anywhere. In fact, you could shuffle both books together, and *still* dip in anywhere. A miscellany, I have said before and stoutly maintain, should be miscellaneous. It need not stick to any fixed logical order. In fact, it *shouldn't*, if only because there isn't one. If I want to sandwich a puzzle allegedly invented by Euclid between a story about Scandinavian kings playing dice for the ownership of an island and a calculation of how likely it is for monkeys to randomly type the complete works of Shakespeare, then why not?

We live in a world where finding time to work systematically through a long and complicated argument or discussion gets ever more difficult. That's still the best way to stay properly informed – I'm not decrying it. I even try it myself when the world lets me. But when the scholarly method won't work, there's an alternative, one that requires only a few minutes here and there. Apparently quite a lot of you find that to your taste, so here we go again. As one radio interviewer remarked about *Cabinet* (sympathetically, I believe), 'I suppose it's the ideal toilet book.' Now, Avril and I actually go out of our way *not* to leave books in the loo for visitors to read, because we don't want to have to bang on the door at 1 a.m. to remove a guest who has found *War and Peace* unexpectedly gripping. And we don't want to risk getting stuck in there ourselves.

But there you go. The interviewer was right. And like its predecessor, *Hoard* is just the kind of book to take on a train, or a plane, or a beach. Or to sample at random over Boxing Day, in between watching the sports channels and the soaps. Or whatever it is that grabs you.

Hoard is supposed to be fun, not work. It isn't an exam, there is no national curriculum, there are no boxes to tick. You don't need to prepare yourself. Just dive in.

A few items do fit naturally into a coherent sequence, so I've put those next to each other, and earlier items do sometimes shed light on later ones. So, if you come across terms that aren't being explained, then probably I discussed them in an earlier item. Unless I didn't think they needed explanation, or forgot. Thumb quickly through the earlier pages seeking insight. If you're lucky, you may even find it.

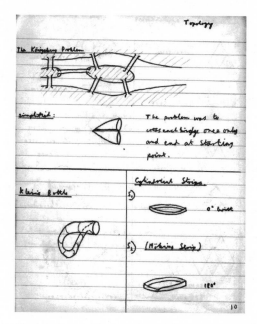

A page from my first notebook of mathematical curiosities.

When I was rummaging through the Cabinet's drawers, choosing new items for my *Hoard*, I privately classified its contents into categories: puzzle, game, buzzword, squib, FAQ, anecdote, infodump, joke, gosh-wow, factoid, curio, paradox, folklore, arcana, and so on. There were subdivisions of puzzles (traditional, logic, geometrical, numerical, etc.) and a lot of the categories overlapped. I did think about attaching symbols to tell you which item is what, but there would be too many symbols. A few pointers, though, may help.

The puzzles can be distinguished from most other things because they end with *Answer on page XXX*. A few puzzles are harder than the rest, but none outlandishly so. The answer is often worth reading even if – especially if – you don't tackle the problem. But you will appreciate the answer better if you have a go at the question, however quickly you give up. Some puzzles are embedded in longer stories; this does not imply that the puzzle is hard, just that I like telling stories.

Almost all the topics are accessible to anyone who did a bit of maths at school and still has some interest in it. The FAQs are explicitly *about* things we did at school. Why don't we add fractions the way we multiply them? What is point nine recurring? People often ask these questions, and this seemed a good place to explain the thinking behind them. Which is not always what you might expect, and in one case not what *I* expected when I started to write that item, thanks to a coincidental email that changed my mind.

However, school mathematics is only a tiny part of a much greater enterprise, which spans millennia of human culture and ranges over the entire planet. Mathematics is essential to virtually everything that affects our lives – mobile phones, medicine, climate change – and it is growing faster than it has ever done before. But most of this activity goes on behind the scenes, and it's all too easy to assume that it's not happening at all. So, in *Hoard* I've devoted a bit more space to quirky or unusual applications of mathematics, both in everyday life and in frontier science. And a bit less to the big problems of pure mathematics, mainly because I covered several of the really juicy ones in *Cabinet*.

These items range from finding the area of an ostrich egg to the puzzling excess of matter over antimatter just after the Big Bang. And I've also included a few historical topics, like Babylonian numerals, the abacus and Egyptian fractions. The history of mathematics goes back at least 5,000 years, and discoveries made in the distant past are still important today, because mathematics builds on its past successes.

A few items are longer than the rest – mini-essays about important topics that you may have come across in the news, like the fourth dimension or symmetry or turning a sphere inside out. These items don't exactly go *beyond* school mathematics: they generally head off in a completely different direction. There is far more to mathematics than most of us realise. I've also deposited a few technical comments in the notes, which are scattered among the answers. These are things I felt needed to be said, and needed to be easy to ignore. I've given cross-references to *Cabinet* where appropriate.

Occasionally you may come across a complicated-looking formula – though most of those have been relegated to the notes at the back of the book. If you hate formulas, *skip these bits*. The formulas are there to show you what they look like, not because you're going to have to pass a test. Some of us *like* formulas – they can be extraordinarily pretty, though they are admittedly an acquired taste. I didn't want to cop out by omitting crucial details; I personally find this very annoying, like the TV programmes that bang on about how exciting some new discovery is, but don't actually tell you anything about it.

Despite its random arrangement, the best way to read *Hoard* is probably to do the obvious: start at the front and work your way towards the back. That way you won't end up reading the same page six times while missing out on something far more interesting. But you should feel positively eager to skip to the next item the moment you feel you've wandered into the wrong drawer by mistake.

This is not the only possible approach. For much of my professional life, I have read mathematics books by starting at the back, thumbing towards the front until I spot something that looks interesting, continuing towards the front until I find the technical terms upon which that thing depends, and then proceeding in the normal front-to-back direction to find out what's really going on.

Well, it works for me. You may prefer a more conventional approach.

Ian Stewart
Coventry, April 2009

A mathematician is a machine for turning coffee into theorems.
Paul Erdős

To Avril, for 40 years
of devotion and support

Calculator Curiosity 1

Get your calculator, and work out:

$(8 \times 8) + 13$

$(8 \times 88) + 13$

$(8 \times 888) + 13$

$(8 \times 8888) + 13$

$(8 \times 88888) + 13$

$(8 \times 888888) + 13$

$(8 \times 8888888) + 13$

$(8 \times 88888888) + 13$

Answers on page 274

Year Turned Upside Down

Some digits look (near enough) the same upside down: 0, 1, 8. Two more come in a pair, each the other one turned upside down: 6, 9. The rest, 2, 3, 4, 5, 7, don't look like digits when you turn them upside down. (Well, you can write 7 with a squiggle and then it looks like 2 upside down, but please don't.) The year 1691 reads the same when you turn it upside down.

Which year in the past is the most recent one that reads the same when you turn it upside down?

Which year in the future is the next one that reads the same when you turn it upside down?

Answers on page 274

Luckless Lovelorn Lilavati

Lilavati.

Among the great mathematicians of ancient India was Bhaskara, 'The Teacher', who was born in 1114. He was really an astronomer: in his culture, mathematics was mainly an astronomical technique. Mathematics appeared in astronomy texts; it was not a separate subject. Among Bhaskara's most famous works is one named *Lilavati*. And thereby hangs a tale.

Fyzi, Court Poet to the Mogul Emperor Akbar, wrote that Lilavati was Bhaskara's daughter. She was of marriageable age, so Bhaskara cast her horoscope to work out the most propitious wedding date. (Right into the Renaissance period, many mathematicians made a good living casting horoscopes.) Bhaskara, clearly a bit of a showman, thought he'd come up with a terrific idea to make his forecast more dramatic. He bored a hole in a cup and floated it in a bowl of water, with everything designed so that the cup would sink when the fateful moment arrived.

Unfortunately, an eager Lilavati was leaning over the bowl, hoping that the cup would sink. A pearl from her dress fell into the cup and blocked the hole. So the cup didn't sink, and poor Lilavati could never get married.

To cheer his daughter up, Bhaskara wrote a mathematics textbook for her.

Hey, thanks, Dad.

Sixteen Matches

Sixteen matches are arranged to form five identical squares.

By moving exactly *two* matches, reduce the number of squares to four. All matches must be used, and every match should be part of one of the squares.

Answer on page 274

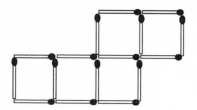

Sixteen matches arranged to form five squares.

Swallowing Elephants

Elephants always wear pink trousers.
Every creature that eats honey can play the bagpipes.
Anything that is easy to swallow eats honey.
No creature that wears pink trousers can play the bagpipes.
Therefore:
Elephants are easy to swallow.

Is the deduction correct, or not?

Answer on page 274

Magic Circle

In the diagram there are three big circles, and each passes through four small circles. Place the numbers 1, 2, 3, 4, 5, 6 in the small circles so that the numbers on each big circle add up to 14.

Answer on page 276

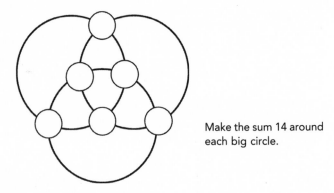

Make the sum 14 around each big circle.

Dodgem

This is a mathematical game with very simple rules that's a lot of fun to play, even on a small board. It was invented by puzzle expert and writer Colin Vout. The picture shows the 4 × 4 case.

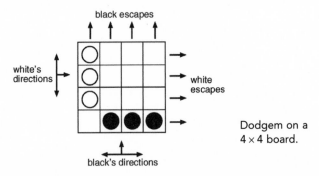

Dodgem on a 4 × 4 board.

Players take turns to move one of their counters one cell

forward, one to the left, or one to the right, as shown by the arrows 'black's directions' and 'white's directions'. They can't move a counter if it is blocked by an opponent's counter or the edge of the board, except for the opposite edge where their counters can escape. A player must always leave his opponent at least one legal move, and loses the game if he does not. The first player to make all his counters escape wins.

On a larger board the initial arrangement is similar, with the lower left-hand corner unoccupied, a row of white counters up the left-hand column, and a row of black counters along the bottom row.

Vout proved that with perfect strategy the first player always wins on a 3×3 board, but for larger boards it seems not to be known who should win. A good way to play is with draughts (checkers) pieces on the usual 8×8 board.

It seems natural to use square boards, but with a rectangular board the player with fewer counters has to move them further, so the game may be playable on rectangular boards. As far as I know, they've not been considered.

• •

Press-the-Digit-ation

I learned this trick from the Great Whodunni, a hitherto obscure stage magician who deserves wider recognition. It's great for parties, and only the mathematicians present will guess how it works.[*] It is designed specifically to be used in the year 2009, but I'll explain how to change it for 2010, and the Answer section on page 276 will extend that to any year.

Whodunni asks for a volunteer from the audience, and his beautiful assistant Grumpelina hands them a calculator. Whodunni then makes a big fuss about it having once been a perfectly ordinary calculator, until it was touched by magic. Now, it can reveal your hidden secrets.

---

[*]   Contrary to widespread belief, mathematicians do go to parties.

'I am going to ask you to do some sums,' he tells them. 'My magic calculator will use the results to display your age and the number of your house.' Then he tells them to perform the following calculations:

- Enter your house number.
- Double it.
- Add 42.
- Multiply by 50.
- Subtract the year of your birth.
- Subtract 50.
- Add the number of birthdays you have had so far this year, i.e. 0 or 1.
- Subtract 42.

'I now predict,' says Whodunni, 'that the last two digits of the result will be your age, and the remaining digits will be the number of your house.'

Let's try it out on the fair Grumpelina, who lives in house number 327. She was born on 31 December 1979, and let's suppose that Whodunni performs his trick on Christmas Day 2009, when she is 29.

- Enter your house number: 327
- Double it: 654.
- Add 42: 696.
- Multiply by 50: 34,800.
- Subtract the year of your birth: 32,821
- Subtract 50: 32,771.
- Add the number of birthdays you have had so far this year (0): 32,771.
- Subtract 42: 32,729.

The last two digits are 29, Grumpelina's age. The others are 327 – her house number.

The trick works for anyone aged 1 to 99, and any house number, however large. You could ask for a phone number instead, and it would still work. But Grumpelina's phone number is ex-directory, so I can't illustrate the trick with that.

If you're trying the trick in 2010, replace the last step by 'subtract 41'.

You don't need a magic calculator, of course: an ordinary one works fine. And you don't need to understand how the trick works to amaze your friends. But for those who'd like to know the secret, I've explained it on page 276.

• •

Secrets of the Abacus

In these days of electronic calculators, the device known as an abacus seems rather outmoded. Most of us encounter it as a child's educational toy, an array of wires with beads that slide up and down to represent numbers. However, there's more to the abacus than that, and such devices are still widely used, mainly in Asia and Africa. For a history, see:
en.wikipedia.org/wiki/Abacus

The basic principle of the abacus is that the number of beads on each wire represents one digit in a sum, and the basic operations of arithmetic can be performed by moving the beads in the right way. A skilled operator can add numbers together as fast as someone using a calculator can type them in, and more complicated things like multiplication are entirely practical.

The Sumerians used a form of abacus around 2500 BC, and the Babylonians probably did too. There is some evidence of the abacus in ancient Egypt, but no images of one have yet been found, only discs that might have been used as counters. The abacus was widely used in the Persian, Greek and Roman civilisations. For a long time, the most efficient design was the one used by the Chinese from the 14th century onwards, called a *suànpán*. It has two rows of beads; those in the lower row signify 1 and those in the upper row signify 5. The beads nearest the dividing line determine the number. The *suànpán* was quite big: about 20 cm (8 inches) high and of varying width depending on the number of columns. It was used lying flat on a table to stop the beads sliding into unwanted positions.

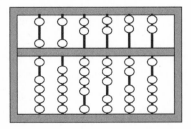

The number 654,321 on a Chinese abacus.

The Japanese imported the Chinese abacus from 1600, improved it to make it smaller and easier to use, and called it the *soroban*. The main differences were that the beads were hexagonal in cross-section, everything was just the right size for fingers to fit, and the abacus was used lying flat. Around 1850 the number of beads in the top row was reduced to one, and around 1930 the number in the bottom row was reduced to four.

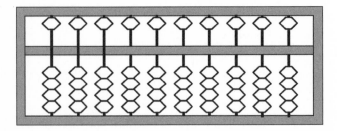

Japanese abacus, cleared.

The first step in any calculation is to clear the abacus, so that it represents $0 \ldots 0$. To do this efficiently, tilt the top edge up so that all the beads slide down. Then lie the abacus flat on the table, and run your finger quickly along from left to right, just above the dividing line, pushing all the top beads up.

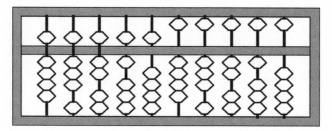

Japanese abacus, representing 9,876,543,210.

Again, numbers in the lower row signify 1 and those in the upper row signify 5. The Japanese designer made the abacus more efficient by removing superfluous beads that provided no new information.

The operator uses the *soroban* by resting the tips of the thumb and index finger lightly on the beads, one either side of the central bar, with the hand hovering over the bottom rows of beads. Various 'moves' must then be learned, and practised, much like a musician learns to play an instrument. These moves are the basic components of an arithmetical calculation, and the calculation itself is rather like playing a short 'tune'. You can find lots of detailed abacus techniques at:

www.webhome.idirect.com/~totton/abacus/pages.htm#Soroban1

I'll mention only the two easiest ones.

A basic rule is: always work from left to right. This is the opposite of what we teach in school arithmetic, where the calculation proceeds from the units to the tens, the hundreds, and so on – right to left. But we *say* the digits in the left–right order: 'three hundred and twenty-one'. It makes good sense to think of them that way, and to calculate that way. The beads act as a memory, too, so that you don't get confused by where to put the 'carry digits'.

To add 572 and 142, for instance, follow the instructions in the pictures. (I've referred to the columns as 1, 2, 3, from the right, because that's the way we think. The fourth column doesn't play any role, but it would do if we were adding, say, 572 and 842, where $5 + 8 = 13$ involves a 'carry digit' 1 in place 4.

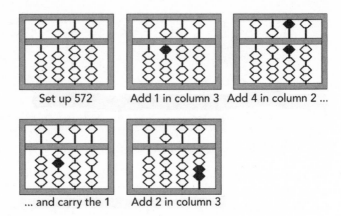

Set up 572   Add 1 in column 3   Add 4 in column 2 ...

... and carry the 1   Add 2 in column 3

A basic technique occurs in subtraction. I won't draw where the beads go, but the principle is this. To subtract 142 from 572, change each digit x in 142 to its *complement* $10 - x$. So 142 becomes 968. Now *add* 968 to 572, as before. The result is 1,540, but of course $572 - 142$ is actually 430. Ah, but I haven't yet mentioned that at each step you subtract 1 from the column one place to the left (doing this as you proceed). So the initial 1 disappears, 5 turns into 4, and 4 turns into 3. The 0 you leave alone.

Why does this work, and why do we leave the units digit unchanged?

Answer on page 277

●●●

Redbeard's Treasure

Captain 'Jolly' Roger Redbeard, the fiercest pirate in the Windlass Islands, stared blankly at a diagram he had drawn in the sand beside the quiet lagoon behind Rope's End Reef. He had buried a hoard of pieces of eight there a few years ago, and now he wanted to retrieve his treasure. But he had forgotten where it was. Fortunately he had set up a cunning mnemonic, to remind him. Unfortunately, it was a bit *too* cunning.

He addressed the band of tattered thugs that constituted his crew.

'Avast, ye stinkin' bilge-rats! Oi, Numbskull, put down that cask o' rotgut and listen!'

The crew eventually quietened down.

'You remember when we boarded the *Spanish Prince*? And just before I fed the prisoners to the sharks, one of 'em told us where they'd hidden their loot? An' we dug it all up and reburied it somewhere safe?'

There was a ragged cry, mostly of agreement.

'Well, the treasure is buried due north o' that skull-shaped rock over there. All we need to know is *how far* north. Now, I 'appens to know that the exact number o' paces is the number of different ways a man can spell out the word TREASURE by puttin' his finger on the T at the top o' this diagram, and then movin' it down one row at a time to a letter that's next to it, one step to the left or right.

'I'm offerin' ten gold doubloons to the first man-jack o' ye to tell me that number. What say ye, lads?'

```
            T
          R   R
        E   E   E
      A   A   A   A
    S   S   S   S
  U   U   U   U   U
R   R   R   R   R   R
E   E   E   E   E   E   E
```

How many paces is it from the rock to the treasure?

Answer on page 277

Hexaflexagons

These are fascinating mathematical toys, originally invented by the prominent mathematician Arthur Stone when he was a graduate student. I'll show you the simplest one, and refer you to the internet for the others.

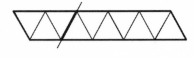

Cut out a strip of 10 equilateral triangles and fold the right-hand end underneath the rest along the solid line . . .

. . . to get this. Now fold the end backwards along the solid line and poke it through . . .

. . . to get this. Finally, fold the grey flap behind and glue it to the adjacent triangle . . .

. . . to get a finished triflexagon.

Having made this curious shape, you can *flex* it. If you pinch together two adjacent triangles separated by a solid line (the edge of the original strip), then a gap opens in the middle and you can flip the edges outwards – turning the hexagon inside out, so to speak. This exposes a different set of faces. It can then be flexed again, which returns it to its starting configuration.

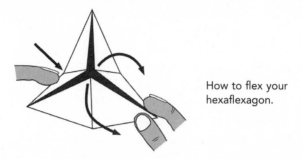

How to flex your hexaflexagon.

All this is easier to do by experimenting on a model than to describe. If you colour the front of the original hexagon red, and the back blue, then the first flex reveals another set of triangles that have not yet been coloured. Colour these yellow. Now each successive flex sends the front colour to the back, makes the back colour disappear, and shows a new colour on the front. So the colours cycle like this:

- Red on front, blue on back
- Yellow on front, red on back
- Blue on front, yellow on back

There are more complicated flexagons, with more hidden faces, which require more colours. Some use squares instead of triangles. Stone formed a 'flexagon committee' with three other graduate students: Richard Feynman, Brent Tuckerman and John Tukey. In 1940, Feynman and Tukey developed a complete mathematical theory characterising all flexagons. A good entry point into the extensive world of the flexagon is en.wikipedia.org/wiki/Flexagon

Who Invented the Equals Sign?

The origins of most mathematical symbols are lost in the mists of antiquity, but we do know where the equals sign = came from. Robert Recorde was a Welsh doctor and mathematician, and in 1557 he wrote *The Whetstone of Witte, whiche is the seconde parte of Arithmeteke: containing the extraction of rootes; the cossike practise, with the rule of equation; and the workes of Surde Nombers.**

In it, he wrote: 'To avoide the tediouse repetition of these woordes: is equalle to: I will sette as I doe often in woorke use, a paire of paralleles, or gemowe† lines of one lengthe: ======, bicause noe .2. thynges, can be moare equalle.'

Robert Recorde and his equals sign.

- -

Stars and Snips

Betsy Ross, who was born in 1752, is generally credited with having sewn the first American flag, with 13 stars representing the 13 founding colonies. (On the present-day Stars and Stripes,

---

* 'Cossike practice' refers to algebra: the Renaissance Italian algebraists referred to the unknown, which we now call x, as *cosa*, Italian for 'thing'. As in *cosa nostra*, 'this thing of ours', referring to the Mafia. 'Surde nombers' are things like square roots, and the word 'surd' still exists in English, though it is seldom used nowadays.

† From the Latin *geminus*, meaning 'twin'.

they are represented by the 13 stripes.) Historians continue to debate the truth of this story, since it is mainly based on word of mouth, and I don't want to get tangled up in the historical arguments: see
www.ushistory.org/betsy/

The important thing for this puzzle is that the stars on the American flag are five-pointed. Apparently George Washington's original design used six-pointed stars, whereas Betsy favoured the five-pointed kind. The committee objected that this type of star was too hard to make. Betsy picked up a piece of paper, folded it, and cut off a perfect five-pointed star with one straight snip of her scissors. The committee, impressed beyond words, caved in.

How did she do that?

Can a similar method make a six-pointed star?

Answers on page 278

Fold and cut this ...   ... to make this.

By the Numbers of Babylon

Ancient cultures wrote numbers in many different ways. The ancient Romans, for instance, used letters: I for 1, V for 5, X for 10, C for 100, and so on. In this kind of system, the bigger the numbers become, the more letters you need. And arithmetic can be tricky: try multiplying MCCXIV by CCCIX, using only pencil and paper.

Our familiar decimal notation is more versatile and better suited to calculation. Instead of inventing new symbols for ever-bigger numbers, it uses a fixed set of symbols, which in Western cultures are 0, 1, 2, 3, 4, 5, 6, 7, 8, 9. Larger numbers are taken care of by using the same symbols in different positions. For instance, 525 means

$$5 \times 100 + 2 \times 10 + 5 \times 1$$

The symbol '5' at the right-hand end stands for 'five'; the same symbol at the left-hand end stands for 'five hundred'. A positional number system like this needs a symbol for zero, otherwise it can't distinguish between numbers like 12, 102, and 1,020.

Our number system is said to be *base 10* or *decimal*, because the value of a digit is multiplied by 10 every time it moves one place to the left. There's no particular mathematical reason for using 10: base 7 or base 42 will work just as well. In fact, any whole number (greater than 1) can be used as a base, though bases greater than 10 require new symbols for the extra digits.

The Mayan civilisation, which goes back to 2000 BC, flourished in Central America from about AD 250 to 900, and then declined, used base 20. So to them, the symbols 5-2-14 meant

$$5 \times 20^2 + 2 \times 20 + 14 \times 1$$

which is 2,054 in our notation. They wrote a dot for 1, a horizontal line for 5, and combined these to get all numbers from 1 to 19. From 36 BC onwards they used a strange oval shape for 0. Then they stacked these 20 'digits' vertically to show successive base-20 digits.

Left: the numbers 0–29 in Mayan; right: Mayan for $5 \times 20^2 + 2 \times 20 + 14 \times 1$

It is often suggested that the Mayans employed base 20 because they counted on their toes as well as their fingers. An alternative explanation occurred to me while I was writing this item. Maybe they counted on fingers and thumbs, with a thumb representing 5. Then each dot is a finger, each bar a thumb, and it can all be done on two hands. Admittedly, we don't have three thumbs, but there are easy ways round this with hands and it's not an issue for symbols. As for the oval shape for zero: don't you agree that it looks a bit like a clenched fist? Meaning no fingers and no thumbs.

This is wild speculation, but I quite like it.

Much earlier, around 3100 BC, the Babylonians had been even more ambitious, using base 60. Babylon is almost a fabled land, with biblical stories of the Tower of Babel and Shadrach in Nebuchadnezzar's furnace, and romantic legends of the Hanging Gardens. But Babylon was a real place, and many of its archaeological remains still survive in Iraq. The word 'Babylonian' is often used interchangeably for several different social groupings that came and went in the area between the Tigris and Euphrates rivers, who shared many aspects of their cultures.

We know a lot about the Babylonians because they wrote on clay tablets, and more than a million of these have survived, often because they were in a building that caught fire and baked the clay rock-hard. The Babylonian scribes used short sticks with shaped ends to make triangular marks, known as cuneiform, in the clay. The surviving clay tablets include everything from household accounts to astronomical tables, and some date back to 3000 BC or earlier.

The Babylonian symbols for numerals were introduced around 3000 BC, and employ two distinct signs for 1 and 10, which were combined in groups to obtain all integers up to 59.

1 𒁹	11 𒌋𒁹	21 𒎙𒁹	31 𒌍𒁹	41 𒃻𒁹	51 𒐐𒁹
2 𒈫	12 𒌋𒈫	22 𒎙𒈫	32 𒌍𒈫	42 𒃻𒈫	52 𒐐𒈫
3 𒐈	13 𒌋𒐈	23 𒎙𒐈	33 𒌍𒐈	43 𒃻𒐈	53 𒐐𒐈
4 𒐉	14 𒌋𒐉	24 𒎙𒐉	34 𒌍𒐉	44 𒃻𒐉	54 𒐐𒐉
5 𒐊	15 𒌋𒐊	25 𒎙𒐊	35 𒌍𒐊	45 𒃻𒐊	55 𒐐𒐊
6 𒐋	16 𒌋𒐋	26 𒎙𒐋	36 𒌍𒐋	46 𒃻𒐋	56 𒐐𒐋
7 𒐌	17 𒌋𒐌	27 𒎙𒐌	37 𒌍𒐌	47 𒃻𒐌	57 𒐐𒐌
8 𒐍	18 𒌋𒐍	28 𒎙𒐍	38 𒌍𒐍	48 𒃻𒐍	58 𒐐𒐍
9 𒐎	19 𒌋𒐎	29 𒎙𒐎	39 𒌍𒐎	49 𒃻𒐎	59 𒐐𒐎
10 𒌋	20 𒎙	30 𒌍	40 𒃻	50 𒐐	

Babylonian numerals from 1 to 59.

The 59 groups act as individual digits in base-60 notation, otherwise known as the sexagesimal system. To save my printer having kittens, I'll do what archaeologists do and write Babylonian numerals like this:

$$5,38,4 = 5 \times 60 \times 60 + 38 \times 60 + 4 = 20,284 \text{ in decimal}$$
$$\text{notation}$$

The Babylonians didn't (until the late period) have a symbol to play the role of our zero, so there was a degree of ambiguity in their system, usually sorted out by the context in which the number showed up. For high precision, they also had a symbol equivalent to our decimal point, a 'sexagesimal point', indicating that the numbers to its right are multiples of $\frac{1}{60}$, $\frac{1}{60} \times \frac{1}{60} = \frac{1}{3600}$, and so on. Archaeologists represent this symbol by a semicolon (;). For example,

$$12, 59; 57, 17 = 12 \times 60 + 59 + \frac{57}{60} + \frac{17}{3600} = 779.955$$

in decimal (to a close approximation).

About 2,000 astronomical tablets have been found, mainly routine tables, eclipse predictions, and so on. Of these, 300 are more ambitious – observations of the motion of Mercury, Mars, Jupiter, and Saturn, for instance. The Babylonians were excellent observers, and their figure for the orbital period of Mars was 12,59;57,17 days – roughly 779.955 days, as we've just seen. The modern figure is 779.936 days.

Traces of sexagesimal arithmetic still linger in our culture. We divide an hour into 60 minutes and a minute into 60 seconds. In angular measure, we divide a degree into 60 minutes and a minute into 60 seconds, too – same words, different context. We use 360 degrees for a full circle, and $360 = 6 \times 60$. In astronomical work, the Babylonians often interpreted the numeral that would usually be multiplied by 60×60 as being multiplied by 6×60 instead. The number 360 may have been a convenient approximation to the number of days in a year, but the Babylonians knew that 365 and a bit was much closer, and they knew how big that bit was.

Nobody really knows why the Babylonians used base 60. The standard explanation is that 60 is the smallest number divisible by 1, 2, 3, 4, 5 and 6. There is no shortage of alternative theories, but little hard evidence. We do know that base-60 originated with the Sumerians, who lived in the same region and sometimes controlled it, but that doesn't help a lot. To find out more, good sites to start from are:

en.wikipedia.org/wiki/Babylonian_numerals

www.gap-system.org/~history/HistTopics/Babylonian_numerals.html

● ●

Magic Hexagons

You've probably heard of magic squares – grids of numbers that add up to the same total when read horizontally, vertically or diagonally. Magic hexagons are similar, but now the grid is a honeycomb, and the three natural directions to read the numbers are at 120° to each other. In *Cabinet* (page 270) I told you that there are only two possible magic hexagons, ignoring symmetrically related ones: a silly one of size 1 and a sensible one of size 3.

The only possible normal magic hexagons, of size 1 and 3, and an abnormal hexagon of size 7.

That's true for 'normal' magic hexagons, where the numbers are consecutive integers starting 1, 2, 3, ... But it turns out that there are more possibilities if you allow 'abnormal' ones, where the numbers remain consecutive but start further along, say 3, 4, 5, ... The largest known abnormal magic hexagon was found by Zahray Arsen in 2006. It has size 7, the numbers run from 2 to 128, and the magic constant – the sum of the numbers in any row or slanting line – is 635. Arsen has also discovered abnormal magic hexagons of size 4 and 5. See en.wikipedia.org/wiki/Magic_hexagon

The Collatz–Syracuse–Ulam Problem

Simple questions need not be easy to answer. Here's a famous example. You can explore it with pencil and paper, or a calculator, but what it does in general baffles even the world's greatest mathematicians. They think they know the answer, but no one can prove it. It goes like this.

Think of a number. Now apply the following rules over and over again:

- If the number is even, divide it by 2.
- If the number is odd, multiply it by 3 and add 1.

What happens?

I thought of 11. This is odd, so the next number is $3 \times 11 + 1 = 34$. That's even, so I divide by 2 to get 17. This is odd, and leads to 52. After that the numbers go 26, 13, 40, 20, 10, 5, 16, 8, 4, 2, 1. From there, we get 4, 2, 1, 4, 2, 1 indefinitely. So usually we add a third rule:

- If you reach 1, stop.

In 1937, Lothar Collatz asked whether this procedure always reaches 1, no matter what number you start with. More than seventy years later, we still don't know the answer. There are several other names for this problem: the Syracuse problem, the $3n + 1$ problem, the Ulam problem. It is often posed as a conjecture which states that the answer is yes, and that's what most mathematicians expect.

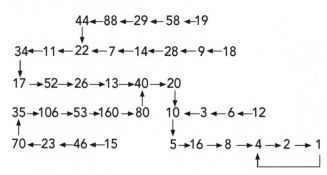

The fates of the numbers 1–20, and anything else they lead to.

One thing that makes the Collatz–Syracuse–Ulam problem or conjecture hard is that the numbers don't always get smaller as you proceed. The chain starting with 15 gets up to 160 before eventually subsiding. Little old 27 positively *explodes*:

$27 \to 82 \to 41 \to 124 \to 62 \to 31 \to 94 \to 47 \to 142 \to 71 \to$
$214 \to 107 \to 322 \to 161 \to 484 \to 242 \to 121 \to 364 \to 182 \to$
$91 \to 274 \to 137 \to 412 \to 206 \to 103 \to 310 \to 155 \to 466 \to$
$233 \to 700 \to 350 \to 175 \to 526 \to 263 \to 790 \to 395 \to 1186 \to$
$593 \to 1780 \to 890 \to 445 \to 1336 \to 668 \to 334 \to 167 \to$
$502 \to 251 \to 754 \to 377 \to 1132 \to 566 \to 283 \to 850 \to 425 \to$
$1276 \to 638 \to 319 \to 958 \to 479 \to 1438 \to 719 \to 2158 \to$
$1079 \to 3238 \to 1619 \to 4858 \to 2429 \to 7288 \to 3644 \to$
$1822 \to 911 \to 2734 \to 1367 \to 4102 \to 2051 \to 6154 \to$
$3077 \to 9232 \to 4616 \to 2308 \to 1154 \to 577 \to 1732 \to 866 \to$
$433 \to 1300 \to 650 \to 325 \to 976 \to 488 \to 244 \to 122 \to 61 \to$
$184 \to 92 \to 46 \to 23 \to 70 \to 35 \to 106 \to 53 \to 160 \to 80 \to$
$40 \to 20 \to 10 \to 5 \to 16 \to 8 \to 4 \to 2 \to 1$

It takes 111 steps to get to 1. But it does get there eventually.

This kind of thing makes you wonder whether there might be some particular number for which the process is even more explosive, and heads off to infinity. The numbers will go up and down a lot, of course. Any odd number leads to an increase, but the number can't increase twice in succession: when n is odd, $3n + 1$ is even, so the next step after that is to divide by 2. But the result at that stage is still bigger than n; in fact, it's $\frac{1}{2}(3n + 1)$. However, if this is also even, we get to something smaller than n, namely $\frac{1}{4}(3n + 1)$. So what happens is quite delicate.

If no number explodes to infinity, the other possibility is that there might be some other cycle which some numbers hit instead of $4 \to 2 \to 1$. It has been proved that any such cycle must contain at least 35,400 terms.

Up to 100 million, the number that takes longest to reach 1 is 63,728,127, which requires 949 steps.

Computer calculations show that every starting number up to at least $19 \times 2^{58} \approx 5.48 \times 10^{18}$ eventually hits 1. This is impressively large, and a lot of theoretical input has to go into the computation – you don't just check the numbers one by one. But the example of Skewes' number (see page 46) shows that 10^{18} isn't really very big, as such things go, so the computer evidence isn't as convincing as it might seem. Everything we know about the question conspires to indicate that if there is an exceptional number that doesn't hit 1, it will be absolutely gigantic.

Probability calculations suggest that the probability that some number heads off to infinity is zero. However, such calculations are not rigorous, because the numbers arising are not truly random. Exceptions might still occur anyway, and even if the argument were rigorous it would not rule out running into a different cycle.

If the process is extended so that it can start with zero or negative integers, four other cycles appear. They all involve numbers bigger than -20, so you might like to search for them (see the answer on page 279). The conjecture now becomes: these five cycles are all that can happen.

There are also connections with chaotic dynamics and fractal geometry, which lead to some beautiful ideas and pictures, but don't solve the problem either. There's a lot of information about this problem on the internet, for example:

en.wikipedia.org/wiki/Collatz_conjecture

mathworld.wolfram.com/CollatzProblem.html

www.numbertheory.org/3x+1/

• •

The Jeweller's Dilemma

Rattler's Jewellers had promised Mrs Jones that they would fit her nine pieces of gold chain together to make a necklace, an endless loop of chain. It would cost them £1 to cut each link, and £2 to rejoin it – a total of £3 per link. If they cut one link at the end of each separate piece, linking the pieces one at a time, the total cost would be £27. However, they had promised to do this for less than the cost of a new chain, which was £26. Help Rattler's avoid losing money – and, more importantly, make the cost to Mrs Jones as small as possible – by finding a better way to fit the pieces of chain together.

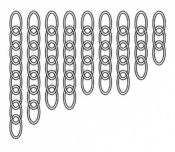

Nine lengths of chain.

Answer on page 279

• •

What Seamus Didn't Know

Our first cat, who rejoiced in the name Seamus Android, was possibly one of the few cats on earth that did *not* always land on its feet. Seamus didn't have a clue. He would come down the stairs one step at a time, head first. At one point, Avril tried to train him to land on his feet by holding him upside down over a thick cushion and letting go. He liked the game but made no effort to turn in mid-air.

Oops ...
What do I do now?

There is a mathematical issue here. Associated with any moving body is a quantity called angular momentum, which, roughly speaking, is the mass multiplied by rate of spin about a suitable axis. Newton's laws of motion imply that the angular momentum of any moving body is conserved, that is, does not change. So how can a falling cat turn over without touching anything?

Answer on page 279

• •

Why Toast Always Falls Buttered-Side Down

A cat is not the only proverbial falling object. Toast is another. It always lands buttered-side down. If not, you must have *buttered the wrong side*.

Curiously, there is some truth to this adage. Robert Matthews has analysed the dynamics of falling toast, which does in fact have a propensity to land in a way that gets butter (or in my case marmalade) all over the carpet and ruins the toast. This lends support to Murphy's law: Anything that can go wrong, will go wrong.

Matthews applied some basic mechanics to explain why toast tends to land buttered-side down. It turns out that tables are just the right height for the toast to make one half turn before it hits the floor. This may not be an accident, because the height of tables is related to the height of humans, and if we were much taller then the force of gravity would smash our skulls if we tripped. Matthews thus traces the trajectory of toast to a universal feature of the fundamental constants of the universe in relation to intelligent life forms. To my mind, this is probably the most convincing example of 'cosmological fine-tuning'.

● ●

The Buttered Cat Paradox

Suppose we put the previous two pieces of folklore together:

● Cats always land on their feet.
● Toast always lands buttered-side down.

Therefore ... what? The buttered cat paradox takes these statements as given, and asks what would happen to a cat, dropped from a considerable height, to whose back is firmly attached a slice of buttered toast – buttered-side outwards from the cat, of course.*

---

* As a practical matter, it is probably a good idea to fit the cat with one of those things that vets use to stop them licking wounds; otherwise the cat will scoff the butter and ruin the experiment.

At the time of writing, the favoured answer is that, as the cat nears the ground, some kind of antigravity effect kicks in, and the cat hovers just off the ground while spinning madly over and over.

However, this argument has some logical loopholes, and it ignores basic mechanics. We've just seen that the mathematics of falling cats, and falling toast, lends scientific support to both adages. So what does the same mathematics say about a buttered cat?

What happens depends on how massive the toast is compared with the cat. If the toast is an ordinary slice, the cat has no difficulty in coping with the small amount of extra angular momentum that the toast contributes, and still lands on its feet. The toast doesn't land at all.

However, if the toast is made of some kind of incredibly dense bread,* so that its mass is much larger than that of the cat, then Matthews's analysis applies and the toast lands buttered-side down with the cat upside down waving its paws frantically in the air.

What happens for intermediate masses? The simplest possibility is that there is a *critical cat-to-toast mass ratio* $[C : T]_{crit}$, below which the toast wins and above which the cat wins. But it wouldn't surprise me to find a range of mass ratios for which the cat lands on its side or, indeed, exhibits more complex transitional behaviour. Chaos cannot be ruled out, as any cat owner knows.

• •

Lincoln's Dog

Abraham Lincoln once asked: 'How many legs will a dog have if you call its tail a leg?'

OK, how many?

Discussion on page 281

• •

* Such as Discworld dwarf bread.

Whodunni's Dice

Grumpelina, the Great Whodunni's beautiful assistant, placed a blindfold over the eyes of the famous stage magician. A member of the audience then rolled three dice.

'Multiply the number on the first dice by 2 and add 5,' said Whodunni. 'Then multiply the result by 5 and add the number on the second dice. Finally, multiply the result by 10 and add the number on the third dice.'

As he spoke, Grumpelina chalked up the sums on a blackboard which was turned to face the audience so that Whodunni could not have seen it, even if the blindfold had been transparent.

'What do you get?' Whodunni asked.

'Seven hundred and sixty-three,' said Grumpelina.

Whodunni made strange passes in the air. 'Then the dice were—'

What? And how did he do it?

Answer on page 282

. . . .

A Flexible Polyhedron

A polyhedron is a solid whose faces are polygons. It has been known since 1813 that a convex polyhedron (one with no indentations) is rigid: it cannot flex without changing the shapes of its faces. This was proved by Augustin-Louis Cauchy. For a long time, no one could decide whether a non-convex polyhedron must also be rigid, but in 1977 Robert Connelly discovered a flexible polyhedron with 18 faces. His construction was gradually simplified by various mathematicians, and Klaus Steffen improved it to a flexible polyhedron with 14 triangular faces. This is known to be the smallest possible number of triangular faces in a flexible polyhedron. You can watch it flex on:

demonstrations.wolfram.com/SteffensFlexiblePolyhedron/
uk.youtube.com/watch?v=OH2kg8zjcqk

You can make one by cutting the diagram from thin card, folding it, and joining the edges marked with the same letters. You can add flaps to do this, or use sticky tape. The dark lines show 'hill' folds, the grey ones 'valley' folds.

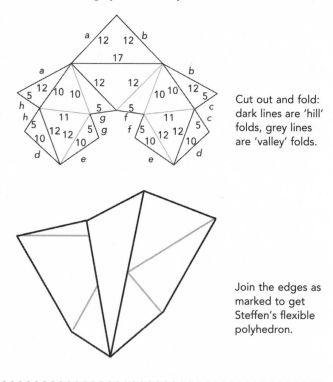

Cut out and fold: dark lines are 'hill' folds, grey lines are 'valley' folds.

Join the edges as marked to get Steffen's flexible polyhedron.

But What About Concertinas?

Hang on a mo – isn't there an obvious way to make a flexible polyhedron? What about the bellows used by blacksmiths to blow air into a fire? Or for that matter, what about a concertina? That has a flexible series of zigzag flaps. If you replace the two big pieces on the ends by flat-sided boxes, which they almost are anyway, then it's a polyhedron. And it's flexible. So what's the big deal?

Although a concertina is a polyhedron, and flexible, it is not a flexible polyhedron. Remember, the shapes of its faces are not permitted to change. They start out flat, so they have to stay flat, which means they can't *bend*. Not even the tiniest bit. But when you play a concertina, and the flexible bit opens up, the faces do bend. Very slightly.

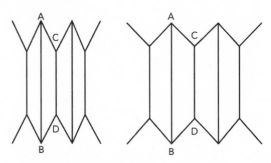

Two positions of a concertina.

Imagine the concertina partially closed, like the left-hand picture, and then opened, like the right-hand one. We're viewing it from the side here. If the faces don't bend or otherwise distort, the line AB can't change length. Now, the sides AC and BD actually slope *away* from us, and we're seeing them sideways on, but, even so, because those lengths don't change in three dimensions, the points C and D in the right-hand picture have to be further apart than they are in the left-hand one. But this contradicts lengths being unchanged. Therefore the faces must change shape. In practice, the material that hinges them together can stretch a bit, which is why a concertina works.

The Bellows Conjecture

Whenever mathematicians make a discovery, they try to push their luck by asking further questions. So when flexible polyhedra were discovered, mathematicians soon realised that there might be another reason why concertinas don't satisfy the mathematical definition. So they did some experiments, making a small hole in a cardboard flexible polyhedron, filling it with smoke, flexing it, and seeing if the smoke puffed out.

It didn't. If you'd done that with a concertina, or bellows, and compressed it, you'd have seen a puff.

Then they did some careful calculations to confirm the experiment, turning it into genuine mathematics. These showed that when you flex one of the known flexible polyhedra, its volume doesn't change. Dennis Sullivan conjectured that the same goes for all flexible polyhedra, and in 1997 Robert Connelly, Idzhad Sabitov and Anke Walz proved he was right.

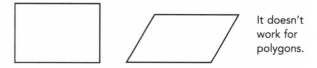

It doesn't work for polygons.

Before sketching what they did, let me put the ideas into context. The corresponding theorem in two dimensions is *false*. If you take a rectangle and flex it to form a parallelogram, the area gets smaller. So there must be some special feature of three-dimensional space that makes a mathematical bellows impossible. Connelly's group suspected it might relate to a formula for the area of a triangle, credited to Heron of Alexandria (see note on page 282).* This formula involves a square root, but it can be rearranged to give a polynomial equation relating the area of the triangle to its three sides. That is, the terms in the equation are powers of the variables, multiplied by numbers.

Sabitov wondered whether there might be a similar equation for *any* polyhedron, relating its volume to the lengths of its sides. This seemed highly unlikely: if there was one, how come the great mathematicians of the past had missed it?

Nevertheless, suppose this unlikely formula *does* exist. Then the bellows conjecture follows immediately. As the polyhedron flexes, the lengths of its sides don't change – so the formula stays

---

* Many historians think that Archimedes got there first.

exactly the same. Now, a polynomial equation may have many solutions, but the volume clearly changes continuously as the polyhedron flexes. The only way to change from one solution of the equation to a different one is to make a jump, and that's not continuous. Therefore the volume cannot change.

All very well, but does such a formula exist? There is one case where it definitely does: a classical formula for the volume of a tetrahedron in terms of its sides. Now, any polyhedron can be built up from tetrahedra, so the volume of the polyhedron is the sum of the volumes of its tetrahedral pieces.

However, that's not good enough. The resulting formula involves all the edges of all the pieces, many of which are 'diagonal' lines that cut across from one corner of the polyhedron to another. These are not edges of the polyhedron, and, for all we know, their lengths may change as the polyhedron flexes. Somehow the formula has to be tinkered with to get rid of these unwanted edges.

A heroic calculation led to the amazing conclusion that such a formula does exist for an octahedron – a solid with eight triangular faces. It involves the 16th power of the volume, not the square. By 1996, Sabitov had found a way to do the same for any polyhedron, but it was very complicated, which may have been why the great mathematicians of earlier times had missed it. In 1997, however, Connelly, Sabitov and Walz found a far simpler approach, and the bellows conjecture became a theorem.

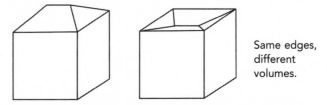

Same edges, different volumes.

I'd better point out that the existence of this formula does not imply that the volume of a polyhedron is uniquely determined by the lengths of its edges. A house with a roof has a smaller volume if you turn the roof upside down. These are two *different* solutions of the same polynomial equation, and that causes no problems in the proof of the bellows conjecture – you can't flex the roof into the downward position without bending something.

● ●

Digital Cubes

The number 153 is equal to the sum of the cubes of its digits:

$$1^3 + 5^3 + 3^3 = 1 + 125 + 27 = 153$$

There are three other 3-digit numbers with the same property, excluding numbers like 001 with a leading zero. Can you find them?

Answer on page 283

● ●

Nothing Which Appeals Much to a Mathematician

In his celebrated book *A Mathematician's Apology* of 1940, the English mathematician Godfrey Harold Hardy had this to say about the digital cubes puzzle:

'This is an odd fact, very suitable for puzzle columns and likely to amuse amateurs, but there is nothing in it which appeals much to a mathematician ... One reason ... is the extreme speciality of both the enunciation and the proof, which is not capable of any significant generalisation.'

In his 1962 book *Profiles of the Future*, Arthur C. Clarke stated three laws about prediction. The first is:

● When a distinguished but elderly scientist states that something is possible, he is almost certainly right. When he states that something is impossible, he is very probably wrong.

This is called Clarke's first law, or often just Clarke's law, and there are good reasons to claim that it applies to Hardy's statement. To be fair, the point Hardy was trying to make is a good one, but you can pretty much guarantee that, whenever anyone cites a specific example to drive such an argument home, it will turn out to be a bad choice. In 2007, a trio of mathematicians – Alf van der Poorten, Kurth Thomsen and Mark Weibe – took an imaginative look at Hardy's assertion. Here's what they found.

It was all triggered by a 'cute observation' made by the number-theorist Hendrik Lenstra:

$$12^2 + 33^2 = 1,233$$

This is about squares, not cubes, but it hints that maybe there is more to this sort of question than first meets the eye. Suppose that a and b are 2-digit numbers, and that

$$a^2 + b^2 = 100a + b$$

which is what you get by stringing the digits of a and b together. Then some algebra shows that

$$(100 - 2a)^2 + (2b - 1)^2 = 10,001$$

So we can find a and b by splitting 10,001 into a sum of two squares. There is an easy way:

$$10,001 = 100^2 + 1^2$$

But 100 has three digits, not two. However, there is also a less obvious way:

$$10,001 = 76^2 + 65^2$$

So $100 - 2a = 76$ and $2b - 1 = 65$. Therefore $a = 12$ and $b = 33$, which leads to Lenstra's observation.

A second solution is hidden here, because we could take $2a - 100 = 76$ instead. Now $a = 88$, and we discover that

$$88^2 + 33^2 = 8,833$$

Similar examples can be found by splitting numbers like 1,000,001 or 100,000,001 into a sum of squares. Number theorists know a general technique for this, based on the prime factors of those numbers. After a lot of detail that I won't go into here, this leads to things like

$$588^2 + 2,353^2 = 5,882,353$$

This is all very well, but what about cubes? Most mathematicians would probably guess that 153 is a special accident. However, it turns out that

$$16^3 + 50^3 + 33^3 = 165,033$$
$$166^3 + 500^3 + 333^3 = 166,500,333$$
$$1,666^3 + 5,000^3 + 3,333^3 = 166,650,003,333$$

and a bit of algebra proves that this pattern continues indefinitely.

These facts depend on our base-10 notation, of course, but that opens up further opportunities: what happens in other number bases?

Hardy was trying to explain a valid point, about what constitutes interesting mathematics, and he plucked the 3-digit puzzle from thin air as an example. If he had given it more thought, he would have realised that although that particular puzzle is special and trivial, it motivates a more general class of puzzles, whose solutions lead to serious and intriguing mathematics.

• •

What Is the Area of an Ostrich Egg?

Who cares, you may ask, and the answer is 'archaeologists'. To be precise, the archaeological team led by Renée Friedman, investigating the ancient Egyptian site of Nekhen, better known by its Greek name Hierakonpolis.

Hierakonpolis was the main centre of Predynastic Egypt, about 5,000 years ago, and it was the cult centre for the falcon-god Horus. It was probably first settled several thousand years earlier. Until recently the site was dismissed as a featureless, barren waste, but beneath the desert sands lie the remains of an ancient town, the earliest known Egyptian temple, a brewery, a potter's house that burnt down when his nearby kiln set it on fire, and the only known burial of an elephant in ancient Egypt.

My wife and I visited this extraordinary site in 2009, under the auspices of the 'Friends of Nekhen'. And there we saw the ostrich eggs whose broken shells were excavated from the area known as HK6. They had been deposited there, intact, as foundation deposits – artefacts deliberately placed in the foundations of a new building. Over the millennia, the eggs had broken into numerous fragments, so the first question was 'how many eggs were there?' The Humpty-Dumpty project – to reassemble the eggs – turned out to be too time-consuming. So the archaeologists settled for an estimate: work out the total area of the shell fragments and divide by the area of a typical ostrich egg.

Typical ostrich egg fragments from Hierakonpolis.

It is here that the mathematics comes in. What is the (surface) area of an ostrich egg? Or, for that matter, what is the area of an egg? Our textbooks list formulas for the areas of spheres, cylinders, cones, and lots of other shapes – but no eggs. Fair enough, since eggs come in many different shapes, but the typical chicken's-egg shape fits ostrich eggs pretty well too, and is one of the commonest shapes found in eggs.

One helpful aspect of eggs is that (to a good approximation, a phrase that you should attach to every statement I make from now on) they are surfaces of revolution. That is, they can be formed by rotating some specific curve around an axis. The curve is a slice through the egg along its longest axis, and has the expected 'oval' shape. The best-known mathematical oval is the ellipse – a circle that has been stretched uniformly in one direction. But eggs aren't ellipses, because one end is more rounded than the other. There are fancier egg-shaped mathematical curves, such as Cartesian ovals, but those don't seem to help.

If you rotate an ellipse about its axis, you get an ellipsoid of revolution. More general ellipsoids do not have circular cross-sections, and are essentially spheres that have been stretched or squashed in three mutually perpendicular directions. Arthur Muir, in charge of the Hierakonpolis eggs, realised that an egg is shaped like two half-ellipsoids joined together. If you can find the surface area of an ellipsoid, you can divide by 2 and then add the areas of the two pieces.

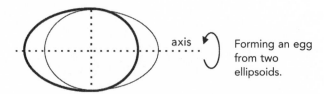

axis

Forming an egg from two ellipsoids.

There is a formula for the area of an ellipsoid, but it involves esoteric quantities called elliptic functions. By a stroke of good fortune, the ostrich's propensity to lay surfaces of revolution, which is a consequence of the tubular geometry of its egg-laying apparatus, comes to the aid of both archaeologist and mathematician. There is a relatively simple formula for the area of an ellipsoid of revolution:

$$A = 2\pi\left(c^2 + ac\,\frac{\arcsin e}{e}\right)$$

where

A = the area

a = half the long axis

c = half the short axis

e = the eccentricity, which equals $\sqrt{1 - c^2/a^2}$

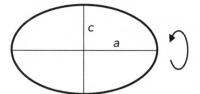

How to rotate the ellipse.

Putting this all together, using measurements from modern ostrich eggs and intact ancient ones, led to an average figure of 570 square centimetres for one egg. This seemed quite large, but experiments with a modern egg confirmed it. The sums then indicated that at least six eggs had been deposited in Structure 07, the largest concentration of ostrich eggs in any single Predynastic deposit.

You never know when mathematics will be useful.

For the archaeological details, see:

www.archaeology.org/interactive/hierakonpolis/field07/6.html

Order into Chaos

Many puzzles, indeed most of them, lead to more serious mathematical ideas as soon as you start to ask more general questions. There is a class of word puzzles in which you have to start with one word and turn it into a different one in such a way that only one letter is changed at each step and that every step is a valid word.* Both words must have the same number of letters, of course. To avoid confusion, you are *not* allowed to rearrange the letters. So CATS can legitimately become BATS, but you can't go from CATS to CAST in one step. You can using more steps, though: CATS–CARS–CART–CAST.

Here are two for you to try:

● Turn SHIP into DOCK.
● Turn ORDER into CHAOS.

Even though these puzzles involve words, with all the accidents and irregularities of linguistic history, they lead to some important and intriguing mathematics. But I'll postpone that until the Answers section, so that I can discuss these two examples there without giving anything away here.

Answers on page 283

● ●

Big Numbers

Big numbers have a definite fascination. The Ancient Egyptian hieroglyph for 'million' is a man with arms outstretched – often likened to a fisherman indicating the size of 'the one that got away', although it is often found as part of a symbolic representation of eternity, with the two hands holding staffs that represent time. In ancient times, a million was pretty big. The Hindu arithmeticians recognised much bigger numbers, and so did Archimedes in *The Sand Reckoner*, in which he estimates how many grains of sand there are on the Earth and demonstrates that the number is finite.

---

* There seems to be no agreed name for such puzzles. 'Change-one-letter-at-a-time-puzzles' is common, but neither concise nor imaginative.

The million that
got away ...

In mathematics and science the usual way to represent big
numbers is to use powers of 10:

$10^2 = 100$ (hundred)
$10^3 = 1,000$ (thousand)
$10^6 = 1,000,000$ (million)
$10^9 = 1,000,000,000$ (billion)
$10^{12} = 1,000,000,000,000$ (trillion)

There was a time when an English billion was 10^{12}, but the
American usage now prevails almost universally – if only because
a billion is now common in financial transactions and we need a
snappy name for it. The obsolete 'milliard' doesn't have the right
ring. In this age of collapsing banks, trillions of pounds or dollars
are starting to be headline material. Billions are passé.

In mathematics, far bigger numbers arise. Not just for the
sake of it, but because they are needed to express significant
discoveries. Two relatively well-known examples are:

$10^{100} = 10,000, \ldots ,000$ (googol)

with one hundred zeros, and

$10^{googol} = 1,000, \ldots ,000$ (googolplex)

which is 1 followed by a googol of zeros. Don't try to write it
down that way: the universe won't last long enough and you
won't be able to get a big enough piece of paper. These two
names were invented in 1938 by Milton Sirotta, the American
mathematician Edward Kasner's nine-year-old nephew, during
an informal discussion of big numbers (*Cabinet*, page 213). The
official name for googol is ten duotrigintillion in the American
system and ten thousand sexdecillion in the obsolescent English

system. The name of the internet search engine Google^{TM} is derived from googol.

Kasner introduced the googol to the world in his book *Mathematics and the Imagination*, written with James Newman, and they tell us that a group of children in a kindergarten worked out that the number of raindrops falling on New York in a century is much less than a googol. They contrast this with the claim (in a 'very distinguished scientific publication') that the number of snowflakes needed to form an ice age is a billion to the billionth power. This is $10^{9000000000}$, and you could just about write it down if you covered every page of every book in a very large library with fine print – all but one symbol being the digit 0. A more reasonable estimate is 10^{30}. This makes the point that it is easy to get confused about big numbers, even when a systematic notation is available.

All of this pales into insignificance when compared with Skewes' number, which is the magnificent

$$10^{10^{10^{34}}}$$

When considering these repeated exponentials, the rule is to start at the top and work backwards. Form the 34th power of 10, then raise 10 to *that* power, and finally raise 10 to the resulting power. A South African mathematician, Stanley Skewes, came across this number in his work on prime numbers. Specifically, there is a well-known estimate for the number of primes $\pi(x)$ less than or equal to any given number x, given by the logarithmic integral

$$\text{Li}(x) = \int_0^x \frac{dt}{\log t}$$

In all cases where $\pi(x)$ can be computed exactly, it is less than Li(x), and mathematicians wondered whether this might always be true. Skewes proved that it is not, giving an indirect argument that it must be false for some x less than his gigantic number, provided that the so-called Riemann hypothesis is true (*Cabinet*, page 215).

To avoid complicated typesetting, and in computer programs, exponentials a^b are often written as $a\char`^b$. Now Skewes' number becomes

$$10\char`^10\char`^10\char`^34$$

In 1955 Skewes introduced a second number, the corresponding one without assuming the Riemann hypothesis, and it is

$$10\char`^10\char`^10\char`^963$$

All this has mainly historical interest, since it is now known that without assuming the Riemann hypothesis, $\pi(x)$ is larger than $\text{Li}(x)$ for some $x < 1.397 \times 10^{316}$. Which is still pretty big.

In our book *The Science of Discworld III: Darwin's Watch*, Terry Pratchett, Jack Cohen and I suggested a simple way to name really big numbers, inspired by the way googol becomes googolplex: If 'umpty' is any number,[*] then 'umptyplex' will mean 10^{umpty}, which is 1 followed by umpty zeros. So 2plex is a hundred, 6plex is a million, 9plex is a billion. A googol is 100plex or 2plexplex, and a googolplex is 100plexplex or 2plexplexplex. Skewes' number is 34plexplexplex.

We decided to introduce this type of name to talk about some of the big numbers appearing in modern physics without putting everyone off. For instance, there are about 118plex protons in the known universe. The physicist Max Tegmark has argued that the universe repeats itself over and over again (including all possible variations) if you go far enough, and estimates that there should be a perfect copy of you no more than 118plexplex metres away. And string theory, the best known attempt to unify relativity and quantum theory, is bedevilled by the existence of 500plex variants on the theory, making it hard to decide which one, if any, is correct.

As far as big numbers go, this is small beer. In my 1969 PhD thesis, in an esoteric and very abstract branch of algebra, I proved that every Lie algebra with a certain property that depends on an

---

[*] It is the favourite number of the Bursar of Unseen University, who is as mad as a hatter.

integer n has another, rather more desirable, property* in which n is replaced by 5plexplexplex...plex with n plexes. I strongly suspected that this could be replaced by $2n$, if not $n + 1$, but as far as I know no one has proved or disproved that, and I've changed my research subject anyway. This tale makes an important point: the usual reason for finding gigantic numbers in mathematics is that some sort of recursive process has been used in a proof, and this probably leads to a wild overestimate.

In orthodox mathematics, the role played by our 'plex' is usually taken over by the exponential function $\exp x = e^x$, and 2plexplexplex will look more like exp exp exp 2. However, 10 is replaced by e here, so this statement is a complete lie. However, it's not hard to complicate it so that it's true, bearing in mind that $e = 10^{0.43}$ or thereabouts. Theorems about repeated exponentials are often rephrased in terms of repeated logarithms, like log log log x (see page 189 for logarithms). For example, it is known that every positive integer, with finitely many exceptions, is a sum of at most

$$n \log n + n \log \log n$$

perfect nth powers – well, ignoring a possible error that is smaller than n. More spectacularly, Carl Pomerance has proved that the number of pairs of amicable numbers (page 110) up to size x is at most

$$x \exp(-c\sqrt{\log \log \log x \log \log \log \log x})$$

for some constant c.

Several systems for representing big numbers have been worked out, with names like Steinhaus–Moser, Knuth's up-arrow and Conway's chained arrow. The topic is much bigger than you

---

* The first property is 'every sub algebra is an n-step subideal', and the second is 'nilpotent of class n'. For example, if every sub-algebra is a 4-step subideal then the algebra is nilpotent of class 5plexplexplexplex, which is bigger than Skewes' number because 5plex is a lot bigger than 34.

might expect, which is only appropriate, and you can find much more about it at

en.wikipedia.org/wiki/Skewes'_number
en.wikipedia.org/wiki/Large_numbers

• •

The Drowning Mathematician

Which (perhaps unfortunately) reminds me:

Q: What sound does a drowning mathematician make?

A: 'log log log log log log log ...'

• •

Mathematical Pirates

Piracy is probably not the first thing that comes to mind in connection with mathematics. Of course, the peak period for piracy, or its state-sanctioned version, privateering, was also the golden age of the mathematics of navigation. Navigators drew geometric diagrams on charts using compasses and protractors; and they 'shot the Sun' with sextants and used mathematical tables to calculate the ship's latitude. But that's not the connection I'm after here, which is a curious set of historical links between mathematicians and pirates, centred on one of the all-time greats: Leonhard Euler, a Swiss-born mathematician who worked in Germany and Russia. He lived between 1707 and 1783 and produced more new mathematics than anybody who has ever lived. The connections were discovered by Ed Sandifer, and posted on his wonderful 'How Euler Did It' website:

www.maa.org/news/howeulerdidit.html

Euler made major advances in mechanics, including extensive applications of the principle of least action, credited to Pierre-Louis Moreau de Maupertuis, an influential French mathematician, writer and philosopher. Maupertuis associated a quantity called 'action' with the motion of any mechanical system, and observed that the actual motion of the system minimises the action, compared with all alternative motions. When a stone bounces down a hill, for instance, the total action is less than it would have been if the stone had started by bouncing uphill for a time, or if it had wandered off sideways, or whatever. Maupertuis was President of the Berlin Academy of Sciences during the period when Euler was in Berlin, and knew Euler well. His father, René Moreau, made the family fortune in the 1690s by attacking British ships, on a privateering licence from the King of France, and married into the aristocracy.

Maupertuis wearing Lapp gear on his 1736 expedition to Lapland, which proved that the Earth is slightly flattened at the poles.

Euler wrote widely about ships,* and in particular analysed their stability, a beautiful application of hydrostatics. His work was not merely theoretical: it had a significant influence on Russian naval shipbuilding. In 1773, he published the *Théorie Complette de la Construction et de la Manoeuvre des Vaissaux Mise à la Portée de Ceux qui s'Appliquent à la Navigation*. In 1776, Henry Watson translated the book into English as *A Complete Theory of the Construction and Properties of Vessels, with Practical Conclusions for the Management of Ships, Made Easy to Navigators*. Watson was a prominent and regular contributor to the *Ladies' Diary*, which

---

* Euler wrote widely about almost everything that had even the slightest connection with mathematics.

contained many mathematical games and problems and was widely read by men as well as women. He borrowed enough money to build three ships, based on some of Euler's work on ship design, and applied to the King of England for a privateer's licence, to operate near the Philippines. When the King declined, Watson used the ships to carry goods instead. Shortly after, he lost £100,000 (the equivalent of about £15–20 million in today's money) on a project to modernise the Calcutta docks for the East India Company. The Company let the project go bankrupt and then bought it for peanuts. On his way back to England to sue the Company, Watson caught a fever and died.

Sir Kenelm Digby was a courtier and diplomat in the reign of King Charles I of England. His link to Euler runs through Fermat, who sent Digby a geometrical problem in 1658. The letter was lost but Digby sent a copy to John Wallis, which has survived. Euler, who made a systematic effort to read everything Fermat wrote, heard of the problem and solved it. Digby had a colourful background. His father, Sir Everard Digby, was executed in 1606 for involvement in the Gunpowder Plot. He dabbled in alchemy, and was a founder of the Royal Society. In 1627–28 he led a privateering expedition to the Mediterranean. Here he seized Spanish, Flemish and Dutch ships, and attacked some French and Venetian ships anchored near the friendly Turkish port of Iskanderun. He returned to England with two ships filled with plunder. However, he also made life difficult for English merchant shipping, by inviting reprisals.

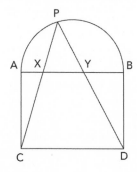

Fermat's poser: Draw a rectangle for which AB is $\sqrt{2}$ times AC, put a semicircle on top, and choose any point P on the semicircle. Construct X and Y as shown. Prove that $AY^2 + BX^2 = AB^2$.

Sandifer also mentions a very tenuous link, through Catherine the Great, who earlier had employed Euler as Court Mathematician, to John Paul Jones, 'Father of the American Navy'. Jones was charged with piracy by the Dutch because he allegedly attacked shipping under 'an unknown flag', but the charge was dropped when the American flag was registered with the appropriate authorities.

The Hairy Ball Theorem

An important theorem in topology says that you can't comb a hairy ball smoothly.* A proof was given in 1912 by Luitzen Brouwer.

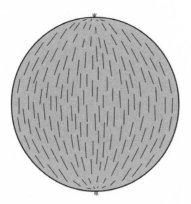

A failed attempt to comb a hairy ball smoothly. At the north and south poles, the hairs would stick up, which is not allowed.

Among the consequences of this theorem is the fact that, at any instant, the horizontal wind speed at some point on the Earth must be zero. Bearing in mind that typical winds are non-zero, such a point will almost always be isolated, and it will often be surrounded by a cyclone. So, at any time, there should be at least one cyclone somewhere in the Earth's atmosphere, for purely topological reasons.

* If that doesn't sound very mathematical, it can be stated more technically: any smooth vector field on a sphere has a singular point. Hope that helps.

The theorem also helps to explain why experimental fusion reactors use toroidal magnetic bottles ('tokamaks') to contain the superheated plasma. You *can* comb a hairy torus (or doughnut*) smoothly. There's more to the physics than that, of course.

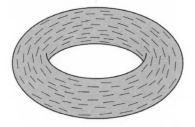

How to comb a hairy doughnut smoothly.

Years ago, one of my mathematical colleagues explained this theorem to a friend of his, and unwisely pointed out that it applied to the family dog. The dog was called 'hairy ball' from that moment on.

The picture shows a combed sphere with two 'tufts' – two places where the hairs don't lay flat. The theorem says there can't be *no* such places, but can there be only one?

Answer on page 287

● ●

Cups and Downs

This puzzle starts with a simple trick involving three cups, which is fun in its own right but also suggests some further questions with surprising answers.

There is a time-honoured way to make money in a pub, requiring three cups and one mug. (The mug is human, and

---

* Throughout this book, 'doughnut' refers to an American one, with a hole. British doughnuts are, or used to be, a single lump, generally filled with jam. Two nations divided by a common culinary heritage. Younger readers may not understand this footnote – all doughnuts have holes, don't they?

should be moderately intoxicated for increased gullibility.) The con-artist places three cups (or glasses) upright on the bar:

He inverts the centre cup

and explains that he will now turn all three of them to the upside-down position in exactly three moves, where each move inverts exactly *two* cups. They need not be adjacent: any two will do. (Of course, this can be done in one move – invert the two end cups – but the requirement to use three moves is part of the misdirection.)

The three moves are:

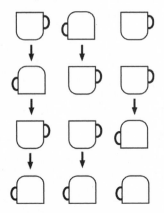

Now the con-artist begins to work on the mug. He casually turns the middle cup upright to get

and invites the mug to repeat the trick, with a small bet on the side to make things more interesting.

Strangely, the cups refuse to behave themselves, no matter what moves the mug attempts. What the mug fails to notice is that the initial position has been changed, surreptitiously. And even if he does notice the change, he may not be aware of the devastating consequences. The parity (odd/even) of the set of upright mugs has now changed from even to odd. But every move *preserves* this parity. The number of upright cups changes by −2, 2 or 0 at each move, so even numbers stay even and odd numbers stay odd. The first starting position has even parity, and so does the required final position. But the second initial position has odd parity. This makes the required final position inaccessible – not just in three moves, but in any number whatsoever.

This disgraceful con-trick (please do *not* try this at home, or in a bar, or anywhere else – or if you do, keep me out of it) shows that there can be obstacles to cup-inversion, but it also misdirects the mug into looking for a three-move solution when the original problem can actually be solved in one move.

The problem can be generalised, with a slight difference from the pub scenario. The resulting puzzle involves the same principles, but it's tidier. I'll ask you two instances of it.

Cups Puzzle 1

Suppose you start with 11 cups, all upside down. The rule is that you must make a series of moves, each of which inverts precisely 4 cups. They do not have to be adjacent. Your objective is to get all 11 cups upright at the same time. Can you do this, and if so, what is the smallest number of moves that does the job?

Cups Puzzle 2

The same question starting with 12 cups, all upside down. Now the rule is that each move must invert precisely 5 cups. Again they do not have to be adjacent. Your have to get all 12 cups upright at the same time. Can you do this, and if so, what is the smallest number of moves required?

Answers on page 287

Secret Codes

Coded messages are as old as writing, but most of the early codes were very easy to break. For instance, the message

QJHT EP OPU IBWF XJOHT

can be decoded as

PIGS DO NOT HAVE WINGS

just by changing each letter into the previous one in the alphabet. If the code shifts the entire alphabet along a number of steps, there are only 25 possibilities to try. Julius Caesar is thought to have used this kind of code, with a shift of 3, in his military campaigns. It has the advantage that encrypting messages (putting them into code) and decrypting them (working out the original 'plaintext' from the coded message) are easy. Its main disadvantage is that you don't have to be very bright to break the code.

You don't have to keep the alphabet in (cyclic) order, of course; you could shuffle it into some random-looking order, giving a *substitution code*. Both sender and recipient must know the shuffled order, so they probably have it written down somewhere, which is potentially insecure. Or else they remember a 'key' such as DANGER! FLYING PIGS, which reminds them to use the order

DANGERFLYIPSBCHJKMOQTUVWXZ

which starts with the letters of the key, ignoring duplicates, and finishes with all the others in alphabetical order. Or maybe reverse alphabetical order, if lots of letters happen to remain unchanged.

Substitution codes are easy to break if the person trying to break them has access to a reasonable quantity of coded messages, because in any language some letters occur more often than others. By calculating the frequency with which each letter occurs – the fraction of times it appears, relative to the total number of letters – you can make an educated guess at the plaintext, and then correct it by looking for words that almost make sense but don't quite.

Typical frequencies of letters in written English.

For instance, in most English writing, the commonest letter is E, followed by T, A, O, I, N, and so on. Of course, texts from different sources may have different frequencies, but all we need is a rough guide. Suppose we already know that in the coded messages, the corresponding 'top six' letters are Z, B, M, X, Q, L. Our first attack on the code message

UXCY RQ LQB KMFZ AXLCY

is to replace each of the letters ZBMXQL by the corresponding ones in ETAOIN. The result (replacing unknown letters by *s) is

*O** *I NIT *A*E *ON**

which doesn't look promising until we realise that NIT is unlikely to appear, whereas NOT is quite plausible. So perhaps X and Q are the wrong way round, and ETAION corresponds to ZBMQXL. Now we decode the message as

*I** *O NOT *A*E *IN**

The second word can't be TO or NO, because T and N have already been used, but it could well be DO. Now we know that letter R encodes D, and so on. If we guess that the twice-used pair CY should be GS, we're looking at

*IGS DO NOT *A*E *INGS

and the code breaks wide open.

What worked (probably badly) for Julius Caesar was not suitable for secure communications in more recent times. Once semaphore, the telegraph and radio were invented, and messages did not have to be carried by a human courier or a pigeon, secure codes became crucial to military and commercial operations. The subjects of cryptography (putting messages into code) and cryptanalysis (decoding them without initially knowing the code) became much more important. Today, almost all countries have big operations in both.

Clearly, the two are linked: in order to break a code you need a lot of sample messages, and some understanding of what sort of code might be involved. Letter frequencies, for instance, do not help much if it is not a substitution code – and it won't be. Each procedure for encrypting messages generates specialised methods for trying to break it.

For very high security, the traditional encryption method is the *one-time pad*. Here, the originator and recipient of the code message both possess a 'notepad' listing 'keys', which are sequences of random numbers. One such sequence is used for any given message, and after use it is destroyed. The numbers on that page are combined with the letters in the plaintext message according to a simple mathematical rule. For instance, successive numbers in the sequence may indicate how far along the alphabet the corresponding letter is to be displaced. So, for example, if the pad reads

5   7   14   22   1   7   16

and the plaintext is

PIGS FLY

then the encrypted message would be

UPUO GSO

(P moves along 5 places, I moves 7 places, and so on). I'm ignoring how to treat spaces, and in practice these should be thought of as additional 'letters'.

The one-time pad was invented in 1917, and it has been proved that any perfectly secure code must use keys that are in some sense equivalent to one-time pad keys. Although one-time pads are secure against any cryptanalysis system, they are not fully secure, because the notepad might be discovered. Originally, the notepad was a physical object – a pad of paper. To reduce the risk of discovery, it was often printed in very small type, and read using a magnifying-glass. For ease of disposal, the pages were made of inflammable material. Today, the 'notepad' might be a computer file.

• •

When 2 + 2 = 0

As a warm-up to more modern methods of data encryption, we need to understand a curious type of arithmetic that goes back to Carl Friedrich Gauss. It is called modular arithmetic, and it is widely used in number theory.

Pick some number, say 4, and call it the modulus. Work only with the whole numbers 0, 1, 2, 3 that lie between 0 (included) and the modulus (excluded). To add two such numbers, add them in the usual way, but if the sum is greater than or equal to 4 (the modulus), subtract a multiple of 4 to reduce the sum to the range 0–3. Do the same for multiplication. So, for example,

$3 + 3 = 6$, subtract 4 (the modulus) to get 2

$3 \times 3 = 9$, subtract 8 (twice the modulus) to get 1.

We can write down addition and multiplication tables:

+	**0**	**1**	**2**	**3**	×	**0**	**1**	**2**	**3**
0	0	1	2	3	**0**	0	0	0	0
1	1	2	3	0	**1**	0	1	2	3
2	2	3	0	1	**2**	0	2	0	2
3	3	0	1	2	**3**	0	3	2	1

Here the boldface numbers show which numbers are being operated on, and the number in the corresponding row and column is the result. For example, to find $3 + 2$ look in row 3, column 2 of the table with a $+$ in the top left-hand corner. The result is 1, so $3 + 2 = 1$.

Well, you may not approve of arithmetic in which $3 + 2 = 1$, but it turns out to be vital to any problem in which what really matters is the *remainder* after dividing by 4. For instance, if you turn some object through four right angles it ends up exactly where it started. So a turn of three right angles followed by two right angles has the same affect as a turn through *one* right angle. (Yes, that's also five right angles, but only 0, 1, 2, 3 right angles are needed to cover all possibilities, so it often makes sense to stay in that range.) So

3 right angles + 2 right angles = 1 right angle

and $3 + 2 = 1$ isn't so silly in this context. Neither is $2 + 2 = 0$, which is how that sum pans out. Rotate through two right angles, and another two right angles, and you get right back where you began – a rotation through zero right angles.

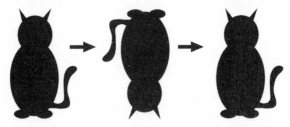

Two right angles plus two right angles equals zero right angles.

The fun comes when you discover that you can use any positive whole number as the modulus – not just 4. The same ideas still work, and they're now sufficiently general to be useful. Any process that repeats the same behaviour over and over again, for instance, may be ripe for analysis using this kind of arithmetic.

When the modulus is 12, we get what is often called clock arithmetic, because on a conventional clock the hour hand gets back to the same position after 12 hours have passed, so multiples of 12 have the same effect as zero.

These funny variants on arithmetic turn up whenever things fit together as part of some cycle that eats its own tail and starts again. It turns out that they obey all of the standard laws of algebra, such as

$$a + b = b + a, \quad ab = ba, \quad a(b + c) = ab + ac$$

and so on. There are a few oddities, though, especially when it comes to division. For example, when working to the modulus 4, the fraction $\frac{1}{2}$ makes no sense. If it did make sense, it would be whichever number, multiplied by 2, gives 1. But the only multiples of 2 are 0 and 2 – the number 1 never appears.

It can be proved that division does make sense whenever the modulus is *prime*, though you still can't divide by 0. For instance, if the modulus is 5, then the two tables above become

+	0	1	2	3	4	×	0	1	2	3	4
0	0	1	2	3	4	**0**	0	0	0	0	0
1	1	2	3	4	0	**1**	0	1	2	3	4
2	2	3	4	0	1	**2**	0	2	4	1	3
3	3	4	0	1	2	**3**	0	3	1	4	2
4	4	0	1	2	3	**4**	0	4	3	2	1

Every number appears in every row of the multiplication table, except row 0, and we can now say things like

$$\frac{3}{2} = 4$$

because

$$2 \times 4 = 3$$

Again, the usual rules for division also work in these cases.

When there is any danger of confusion, mathematicians write these equations like this:

$$2 \times 4 \equiv 3 \quad (\text{mod } 5)$$

with a special symbol $\equiv$ replacing the equals sign, and a reminder of which modulus is involved, to make it clear that they don't really think that $2 \times 4 = 3$. But often they don't bother.

•••

Secret Codes That Can Be Made Public

Arithmetic to a modulus is the key (no pun intended) to a remarkable development in cryptography: the public key cryptosystem. All codes rely on secret keys, and the biggest danger is that an eavesdropper finds out what the key is. If the enemy gets hold of a copy of your one-time pad, perhaps through the actions of a spy, you're in deep trouble.

Or maybe not. The tacit assumption here is that, once someone knows the key, they can easily decode the message. After all, that's what the intended recipient has to do, so it's silly to make it hard. But in 1977 Ron Rivest, Adi Shamir and Leonard Adleman discovered that matters aren't quite so straightforward. In fact, it is possible to make public the key for putting a message into code, without anyone being able to deduce the inverse procedure of decoding the message. However, the legitimate recipient can decode the message using a different, related key – which is kept secret.

Methods like this rely on a curious mathematical fact: reversing a calculation, working back from the answer to the question, can sometimes be much harder than doing the calculation itself – even when the process is in principle reversible.* If so, knowing the procedure concerned does not make it possible to work out how to undo it. But this fact alone is no use unless there is some secret short cut, so that the intended recipient can undo the encoding procedure easily. And it is here that arithmetic to a modulus, Gauss's bizarre invention in which $2 + 2$ might be 0, comes into its own.

The RSA cryptosystem, named after the initials of its aforementioned inventors, is based on a theorem proved by Euler, generalising a simpler one discovered and proved by Pierre de Fermat. The simpler version is called Fermat's Little Theorem, to distinguish it from his Last (or 'Great') Theorem (*Cabinet,* page 50). It states that if you are working to a prime modulus p, then $a^{p-1} \equiv 1$ for any number a. For example, with 5 as the modulus, we should find that $1^4 \equiv 1$, $2^4 \equiv 1$, $3^4 \equiv 1$, $4^4 \equiv 1$. And they are. For instance,

$$3^4 \equiv 3 \times 3 \times 3 \times 3 \equiv 81 \equiv 1 \quad (\text{mod } 5)$$

because $81 - 1 = 80$, which is divisible by 5. The same sort of thing works for the other numbers.†

To apply RSA encryption, you first represent messages by numbers. For example, every block of 100 letters, spaces and other characters could be represented as a 200-digit number,

---

*   Such procedures are often likened to trapdoors, where it is easy to go in but hard to get out. I'm inclined to compare them to catflaps. Our cat Harlequin knows how to go out of a catflap, by pushing, but most of the time she imagines that the way to get back in is to reverse the procedure, and sit outside trying to pull the flap open. It wouldn't surprise me if she took that to the logical extreme and tried to come in tail first. She forgets the secret short cut, and we lie in bed listening to the racket thinking: 'Harley! *Push!*'

†   Fermat proved this theorem before Gauss invented modular arithmetic, but not from that point of view.

where each successive pair of digits encodes characters according to the rule A = 01, B = 02, ..., Z = 26, [space] = 27, ? = 28, and so on. Then a message breaks up into a series of 100-digit numbers. Let N be one of those numbers. Our first task is to put N into code, and we do that using a mathematical recipe in modular arithmetic.

I'll start with an example, using numbers much smaller than those used in practice.

Alice uses two special numbers: 77 and 13, which can be made public. Suppose that her message is $N = 20$. Then she calculates 20^{13} (mod 77), which is 69, and sends that to Bob.

Bob knows the secret number 37, which reverses what Alice does with 13. He decodes Alice's message by raising it to that power (mod 77):

$$69^{37} \equiv 20 \quad (\text{mod } 77)$$

This works for any message that Alice might send, because

$$(N^{13})^{37} \equiv N \quad (\text{mod } 77)$$

Where do these numbers come from?

Alice's choice of 77 is the product of two primes, 7×11. Euler's theorem applies to the number $(7 - 1) \times (11 - 1)$, which is 60. It tells us that there is some number d such that $13^d \equiv 1$ (mod 60), and then $(N^{13})^d \equiv N$ (mod 77) for any message N. As Bob – and only Bob – knows, $d = 37$.

To make this method practical, we replace 7 and 11 by much larger primes – typically with 100 digits or thereabouts (see note on page 288). The encoding key (here 13) and decoding key (here 37) can be calculated from those primes. Only the encoding key and the product of the two primes, a 200-digit number, need be made public. Only Bob need know the decoding key.

This involves finding really big primes, which turns out to be easier than we might expect: there are efficient ways to test whether a number is prime without looking for factors. And, of course, you have to use a computer to do the sums. Note the catflap effect: Alice doesn't need to know how to decode messages, only how to encode them. Mathematicians generally think, but can't yet prove, that working out the prime factors of a really big number is extremely hard – so hard that in practice it can't be done, no matter how big and fast your computer might be. *Finding* big primes is much easier, and so is multiplying them together.

Of course, in my example, with impractically small numbers, finding the decoding key 37 is easy. Alice could work it out, and so could any eavesdropper. But with 100-digit primes, say, calculating the decoding key seems to be impossible if all you know is the product of the two primes. On the other hand, if you do know the primes, then it is relatively straightforward to find the decoding key. That's why it's possible to set up the system to begin with.

Systems like RSA are very suitable for the internet, where every user has to 'know' how to send an encrypted message (such as a credit card number). That is, the method for encrypting this message has to be stored on their computer. So a skilled programmer could find it. But only the bank needs to know the decryption key. So until criminals discover efficient ways to factorise big numbers into primes, your money is safe. Assuming it's safe in the hands of the banks, which has suddenly become questionable.

In practical applications, some care has to be taken and the method isn't quite this simple. See, for example: en.wikipedia.org/wiki/RSA

It is also worth remarking that, in practice, RSA is mainly used for sending encrypted versions of *keys* to other, simpler cryptosystems, which can then be used to send messages, rather than using RSA for the messages themselves. RSA involves a bit too much computational time to be used routinely for messages.

There is a curious historical postscript to this story. In 1973, the same method was invented by Clifford Cocks, a mathematician working for British Intelligence, but it was considered impractical at the time. Because his work was classified Top Secret, no one knew about his anticipation of the RSA system until 1997.

• •

Calendar Magic

'My beautiful assistant,' stated the Great Whodunni, 'will now hand me a perfectly ordinary calendar.'

Grumpelina smiled sweetly and did as instructed. It was indeed an ordinary calendar, with seven columns per month, headed by the days Sunday–Saturday, and the numbers of the days written in order.

Whodunni then called for a volunteer from the audience, while Grumpelina blindfolded him (Whodunni, that is).

'I want you to choose any month from the calendar, and then draw a 3×3 square round nine dates. Don't include any blank spaces. I will then ask you to tell me the smallest number from those dates, and I will *instantly* tell you what the nine numbers add up to.'

The volunteer did so, and, as it happened, he chose a square of dates for which the smallest number was 11. As soon as he told the magician this number, Whodunni immediately replied '171'.

Whodunni's method works whichever 3×3 square is chosen. How does he do it?

Answer on page 289

10	11	12	13	14
17	18	19	20	21
24	25	26	27	28

The volunteer's choice.

• •

Mathematical Cats

Isaac Newton, it is said,* had a cat. He cut a hole in the bottom of his study door so that puss could get in and out. So we must add to Newton's achievements the invention of the catflap, except that his version lacked the flap. Anyway, the cat had kittens. So Newton cut a small hole in the door next to the bigger one.

I don't know whether Lewis Carroll – pen-name of the mathematician Charles Lutwidge Dodgson – had a cat, but he created one of the most memorable fictional cats: the Cheshire Cat, which slowly faded away until only its grin remained. The Cheshire isn't a breed of cat: it is an English county where cheese was (and still is) made. Possibly Carroll was referring to the British shorthair, a breed of cat that appeared on Cheshire Cheese labels.

The Cheshire Cat.

Problem 79 of the ancient Egyptian Rhind Papyrus (pages 77–8) poses the sum

houses	7
cats	49
mice	343
wheat seed	2,401
hekat	16,807 (a *hekat* is a measure of volume)
TOTAL	19,607

---

* This is the time-honoured formula for: 'Somebody told me this, but I can't provide a shred of evidence.'

where each number is 7 times the previous one. The scribe gives a short cut:

$$2{,}801 \times 7 = 19{,}607$$

Note that $2{,}801 = 1 + 7 + 49 + 343 + 2{,}401$. These numbers are the first few powers of 7. I have no idea why the scribe thought it sensible to add up such diverse items, mind you.

Still on exponential growth: the Humane Association has pointed out that if two cats and their kittens breed for 10 years, with each cat having two litters of three surviving kittens per year, then the cat population grows like this:

> 12   66   382   2,201   12,680   73,041   420,715
> 2,423,316   13,968,290   80,399,780

In the 1960s the Russian mathematician Vladimir Arnold studied a map (another word for 'function' or 'transformation') from the torus to itself, defined by

$$(x, y) \rightarrow (2x + y, x + y) \quad (\text{mod } 1)$$

where x and y lie between 0 (included) and 1 (excluded), and (mod 1) means that everything before the decimal point (the integer part) is ignored. So 17.443 (mod 1) = 0.443, for instance. The dynamics of this map are chaotic (*Cabinet*, page 117); also, it 'preserves area', meaning that areas don't change when it is applied. So it provided a simple model for more complicated area-preserving maps arising naturally in mechanics.

This map quickly became known as Arnold's cat, because he illustrated its effect by drawing a cat on the torus, and showing how the cat distorts when the map is applied. The same thing is done with a picture of a real cat at:
upload.wikimedia.org/wikipedia/commons/a/a6/Arnold_cat.png
www.nbi.dk/CATS/PICS/cat_arnold.gif

Author Theoni Pappas wrote a children's book, *The Adventures of Penrose the Mathematical Cat*, presumably named after mathematical physicist Roger Penrose.

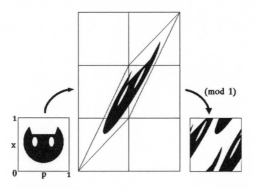

Arnold's cat.

In the book *Mathematicians in Love* by Rudy Rucker, two mathematics graduate students prove a theorem characterising all dynamical systems in terms of objects from Dr Seuss's *The Cat in the Hat*.[*]

In his 1964 research text *Abelian Categories*, Peter Freyd included the index entry 'kittygory'. The page concerned refers to a 'small category'.

There's a mathematician named Nicholas Katz – does that count?

Um – Felix Hausdorff?

• •

---

[*] When I was doing a lecture tour of Oregon, I once stayed in the Sylvia Beach Hotel, whose rooms have literary themes: the Oscar Wilde room, the Agatha Christie room. Mine was the Dr Seuss room, with a 15-foot (5-metre) Cat in the Hat painted on one wall.

The Rule of Eleven

There's an old test for divisibility by 11, seldom taught in these days of calculators. Suppose, for example, that the number is 4,375,327. Form the two sums

$$4 + 7 + 3 + 7 = 21, \qquad 3 + 5 + 2 = 10$$

formed by taking every alternate digit (**4**3**7**5**3**2**7**). Take the difference, $21 - 10 = 11$. If this difference is exactly divisible by 11, so is the original number, and conversely. (The number 0 is exactly divisible by 11, being equal to 11×0.) Here the difference is 11 itself, which is divisible by 11, so the test says that 4,375,327 is divisible by 11. In fact, it is equal to $11 \times 397,757$. Initial zeros make no difference, by the way, since they add zero to whichever sum they appear in.

Here are two puzzles and a question; the puzzles are easier if you use this test.

- Find the largest number that uses each of the digits 0–9 exactly once, and is divisible by 11 without remainder.
- Find the smallest such number, not starting with 0.
- While we're at it: what is the smallest positive multiple of 11 for which the test does not yield a difference of zero?

Answers on page 290

Digital Multiplication

The square array

1	9	2
3	8	4
5	7	6

uses each of the nine digits 1–9. The second row 384 is twice the first row 192, and the third row 576 is three times the first row.

There are three other ways to do this. Can you find them?

Answer on page 291

Common Knowledge

There is an entire genre of puzzles that rests on the counterintuitive properties of 'common knowledge' – something that has been made public, so that not only does everyone involved know it, but they know that everyone knows it, *and* they know that everyone knows that everyone knows it ... A traditional case concerns the curious habits of the obscure but very polite Glaberine* order of monks.

Not 'habits' as in clothing, you appreciate.

Brothers Aelfred, Benedict and Cyril are asleep in their cell, when the novice Legpulla sneaks in and paints a blue blob on the top of each of their shaven heads. When they awake, each notices the blob on the other's head. Now, the monastery rules are clear: it is impolite to say anything that will cause direct embarrassment to another member of the order, but it is also impolite to conceal anything embarrassing about *yourself*. And impoliteness is not permitted under any circumstances. So the monks say nothing, and their demeanour gives no hint of what they have seen.

Each vaguely wonders whether he, too, has a blob, but dare not ask, and there are no mirrors in their cell, nothing reflective at all. And so things remain until the Abbot enters, frowns, and informs them (neatly avoiding *direct* embarrassment) that 'At least one of you has a blue blob on his head.'

Of course, all three monks know that. So does the information make any difference to them?

If you've not met this puzzle before, it helps to start with a simpler version, just two monks, Aelfred and Benedict. Each can see the other's blob, but has no idea what his own head might bear. After the Abbot's public announcement, Aelfred starts thinking. '*I* know Benedict has a blob, but *he* doesn't, because he can't see the top of his own head. Dear Lord, do *I* have a blob? Hmmm ... Suppose I *don't* have a blob. Then Benedict will see

---

* Look up *glaber* in Latin.

that I don't, so he will immediately deduce from the Abbot's remark that *he* must have a blob. But he hasn't shown any sign of embarrassment. Oh dear, *I* must have a blob.' Benedict comes to a similar conclusion.

Without the Abbot's remark, these deductions don't work, yet the Abbot tells them nothing – apparently – that they don't know already. Except ... Each knew that at least one monk (the other one) had a blob, but they didn't know that the *other* monk knew that at least one monk had a blob.

Got that? Very well – what happens with three monks? Again, they can all deduce that they have blobs, but only after the Abbot's announcement (see the answers on page 291). The same goes when there are four, five, or more monks, if all of them have blobs on their heads. Indeed, suppose there are 100 monks. Each bears a blob, each is unaware of that, and each is an amazingly rapid logician. To avoid timing issues, suppose that the Abbot has a bell. 'Every ten seconds,' he tells them, 'I will ring this bell. That will give you time to carry out the necessary logic. Immediately after I ring, all monks who can deduce logically that they have a blob must put their hands up.' He waits ten minutes, ringing his bell from time to time, but nothing happens. 'Oh, yes, I forgot,' he says. 'Here is one extra piece of information. At least one of you has a blob.'

Now nothing happens for 99 rings, and then all 100 monks simultaneously raise their hands after the 100th ring.

Why? Monk number 100, say, can see that the other 99 all have blobs. 'If I do not have a blob,' he thinks, 'then the other 99 all know this. That takes me out of the reckoning altogether. So they are making whatever series of deductions you get with 99 monks when I don't have a blob. If I've sorted out the 99-monk logic right, then after 99 rings they will all put up their hands.' He waits for ring 99, and nothing happens. 'Ah, so my assumption is wrong, and I must have a blob.' Ring 100, up goes his hand. Ditto for the other 99 monks.

Ah, yes ... but maybe monk 100 was wrong about the 99-monk logic. Then it all falls apart. However, the 99-monk logic (with the hypothetical assumption that monk 100 is blobless) is the same. Now monk 99 expects the other 98 to put up their hands at the 98th ring, *unless* monk 99 has a blob. And so it goes, recursively, until we finally get down to a single hypothetical monk. He sees no blobs anywhere, is startled to discover that somebody has one, immediately deduces it must be *him* (you don't need to be an expert logician at that stage) and puts his hand up after the first ring.

Since his 1-monk logic is correct, so is the 2-monk logic, then the 3-monk ... all the way to the 100-monk logic. So this puzzle is a striking example of the Principle of Mathematical Induction. This says that if some property of whole numbers holds for the number 1, and if its validity for any given number implies its validity for the next number, no matter what those numbers may be, then it must be valid for *all* numbers.

That's the usual story, but there's more. So far I've assumed that every monk has a blob. However, very similar reasoning shows that this requirement is not essential. Suppose, for example, that 76 monks out of a total of 100 have blobs. Then, if everyone is logical, nothing happens until just after the 76th ring, when all the monks with blobs put up their hands simultaneously, but none of the others.

At first sight, it's hard to see how they can work this out. The trick lies in the synchronisation of their deductions by the bell, and the application of common kowledge. Try two or three monks first, with different numbers of blobs, or cheat by peeking at the answers on page 291.

● ●

Pickled Onion Puzzle

Three weary travellers came to an inn, late in the evening, and asked the landlord to prepare some food.

'All I got is pickled onions,' he muttered.

The travellers replied that pickled onions would be fine, thank you very much, since the alternative was no food at all.

The landlord disappeared and eventually came back with a jar of pickled onions. By then, all the travellers had fallen asleep, so he put the jar on the table and went off to bed, leaving his guests to sort themselves out.

The first traveller awoke. Not wishing to make a pig of himself, and not knowing what anyone else had already eaten, he took the lid off the jar, threw away an onion that looked bad, ate one-third of the onions that remained, put the lid back on the jar, and went back to sleep.

The second traveller awoke. Not wishing to make a pig of himself, and not knowing what anyone else had already eaten, he took the lid off the jar, threw away two onions that looked bad, ate one-third of the onions that remained, put the lid back on the jar, and went back to sleep.

The third traveller awoke. Not wishing to make a pig of himself, and not knowing what anyone else had already eaten, he took the lid off the jar, threw away three onions that looked bad, ate one-third of the onions that remained, put the lid back on the jar, and went back to sleep.

At this point the landlord returned and removed the jar, which now contained six pickled onions.

How many were there to start with?

Answer on page 292

. . . .

Guess the Card

The Great Whodunni has an endless supply of mathematical card tricks. This one allows him to identify a specific card, chosen from 27 cards taken from a standard pack.

Whodunni shuffles the 27 cards, and lays them out in a fan so that his victim can see all of them.

'Choose one card, mentally, and remember it,' he tells him. 'Turn your back, write down the card, and seal it in this envelope, so that we can verify your choice at the end.'

Now Whodunni deals out the 27 cards, face up, into three piles of 9 cards each, and asks the victim to say which pile the chosen card is in.

He picks up the piles, stacks them together without shuffling, then deals them into three piles and asks for the same information.

Finally, he picks up the piles, stacks them together without shuffling, then deals them into three piles and asks for the same information for a third time.

Then he picks out the chosen card.

How does the trick work?

Answer on page 293

And Now with a Complete Pack

Whodunni can do even better. In just two deals, he can correctly identify a card chosen from the full 52-card pack.

First, he deals the cards in 13 rows of 4 cards, and asks which row the card is in.

Then he reassembles the cards into the same order, deals them into 4 rows of 13 cards, and again asks which row the card is in.

After which he unerringly names the chosen card.

How does this trick work?

Answer on page 293

Halloween = Christmas

Why do mathematicians always confuse Halloween and Christmas?

Answer on page 293

Egyptian Fractions

Whole numbers are fine for addition and multiplication, but subtraction causes problems because, for instance, $6 - 7$ doesn't work with positive whole numbers. This is why negative numbers were invented. A positive or negative whole number is called an integer.

In the same way, the problem of dividing one number by another, such as $6 \div 7$,[*] requires the invention of fractions like $\frac{6}{7}$. The number on the top (here 6) is the *numerator*, the one on the bottom (here 7) is the *denominator*.

Historically, different cultures handled fractions in different ways. The ancient Egyptians had a very unusual approach to fractions; in fact, they had *three* unusual approaches.

First, they had special hieroglyphs for $\frac{2}{3}$ and $\frac{3}{4}$.

Hieroglyphs for $\frac{2}{3}$ and $\frac{3}{4}$.

Second, they used various portions of the Eye of Horus, or Wadjet Eye, to represent 1 divided by the first six powers of 2.

Wadjet Eye (left), and fraction hieroglyphs derived from it (right).

Finally, they devised symbols for fractions of the form 'one over something', that is, $\frac{1}{2}, \frac{1}{3}, \frac{1}{4}, \frac{1}{5}$, and so on. Today we call these

---

[*] This is the only time that the old-fashioned 'division' symbol $\div$ will appear in this book. Oops.

unit fractions. The unit fraction $1/n$ was represented by placing a cushion-shaped hieroglyph (normally representing the letter R) over the top of the symbols for n.

Hieroglyphs for 1/1,237 (in practice the Egyptians wouldn't have used such big numbers in a unit fraction).

However, these methods dealt only with special types of fractions, and 6 divided by 7 was still a problem. So the Egyptians expressed all other fractions as sums of *distinct* unit fractions, for instance

$$\frac{2}{3} = \frac{1}{2} + \frac{1}{6}$$

and

$$\frac{6}{7} = \frac{1}{2} + \frac{1}{3} + \frac{1}{42}$$

It's not at all clear why they didn't like to write $\frac{2}{3}$ as $\frac{1}{3} + \frac{1}{3}$, but they didn't.

Doing arithmetic with unit fractions is weird, but possible. Our method is very different: we 'put both fractions over a common denominator' (page 310) like this:

$$\frac{2}{3} + \frac{6}{7} = \frac{2 \times 7 + 6 \times 3}{3 \times 7} = \frac{14 + 18}{21} = \frac{32}{21} = 1\frac{11}{21}$$

We can see that the result is roughly $1\frac{1}{2}$, which isn't obvious from the Egyptian fractions.

Nevertheless, the Egyptians did amazing things with their symbolism. Our most important source for their work is the Rhind mathematical papyrus, now in the British Museum. Alexander Rhind bought the papyrus in 1858 in Luxor; it seems to have been unearthed by unauthorised excavations near the Ramesseum.

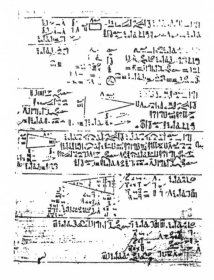

Part of the Rhind mathematical papyrus.

The papyrus dates to around 1650 BC, in the Second Intermediate Period. The scribe Ahmose copied it from an earlier text from the time of the 12th dynasty pharaoh Amenemhat III, two centuries earlier, but the original text is lost. It measures 33 cm by 5 m, and even now scholars do not understand everything on it. However, one remarkable section, about one-third of one side, deals with unit-fraction representations of numbers of the form $2/n$, where n is odd and runs from 3 to 101.

Ahmose's results here can be summed up in a table. To simplify the printing and improve legibility, an entry like

17   12   51   68

means that

$$\frac{2}{17} = \frac{1}{12} + \frac{1}{51} + \frac{1}{68}$$

The table is impressive, but also raises a number of questions. How did whoever first found these representations discover them? Why did the scribes prefer these particular representations?

Expressing 2/n, for n odd, as a sum of at most four unit fractions.

n	2/n in unit fractions	n	2/n in unit fractions
3	special hieroglyph	53	30 318 795
5	3 15	55	30 330
7	4 28	57	38 114
9	6 18	59	36 236 531
11	6 66	61	40 244 488 610
13	8 52 104	63	42 126
15	10 30	65	39 135
17	12 51 68	67	40 335 536
19	12 76 114	69	46 138
21	14 42	71	40 568 710
23	12 276	73	60 219 292 365
25	15 75	75	50 150
27	18 54	77	44 308
29	24 58 174 232	79	60 237 316 790
31	20 124 155	81	54 162
33	22 66	83	60 332 415 498
35	30 42	85	51 255
37	24 111 296	87	58 174
39	26 78	89	60 356 534 890
41	24 246 328	91	70 130
43	42 86 129 301	93	62 186
45	30 90	95	60 380 570
47	30 141 470	97	56 679 776
49	28 196	99	66 198
51	34 102	101	101 202 303 606

In 1967, at the request of Richard Gillings, C. L. Hamblin programmed an early electronic computer belonging to Sydney University to list all possible ways to represent the fractions 2/n in Ahmose's table as sums of unit fractions. The results led Gillings to argue that:

- the Egyptians preferred small numbers;

- they preferred sums with two unit fractions to those with three, and sums with three unit fractions to those with four;
- usually they liked the first number to be as small as possible, but not when that made the last number too big;
- they preferred even numbers, even when this led to bigger numbers or more of them.

For example, the computer found that

$$\frac{2}{37} = \frac{1}{19} + \frac{1}{703}$$

but both numbers are odd and 703 is large. The scribes preferred

$$\frac{2}{37} = \frac{1}{24} + \frac{1}{111} + \frac{1}{296}$$

with two even numbers and nothing very big. Gillings gives an extensive discussion in his *Mathematics in the Time of the Pharaohs*. This book is a bit long in the tooth, and the historical study of Egyptian mathematics has moved on, but it still has a lot of interesting things to say.

● ●

The Greedy Algorithm

Egyptian fractions are obsolete for practical arithmetic, but still very much alive as mathematics, and you can learn a lot about modern fractions by thinking about Egyptian ones. For a start, it's not obvious that every fraction less than one *has* an 'Egyptian representation' – as a sum of distinct unit fractions – but it's true. Leonardo of Pisa, the famous 'Fibonacci' (*Cabinet*, page 98), proved this in 1202, showing that what is nowadays called the 'greedy algorithm' always does the job. An *algorithm* is a specific method of calculation that always produces an answer, like a computer program.

The greedy algorithm begins by finding the largest unit fraction that is less than or equal to the fraction you want to represent – that's what makes it greedy. Subtract this fraction from the original fraction. Now repeat, looking for the largest unit fraction that is *different* from the one you got the first time, but less than what's left. Keep going.

Amazingly, this method eventually reaches a unit fraction and stops.

Let's try out the greedy algorithm on the fraction $\frac{6}{7}$.

- Find the biggest unit fraction that is less than or equal to $\frac{6}{7}$. This is $\frac{1}{2}$.
- Find the difference $\frac{6}{7} - \frac{1}{2} = \frac{5}{14}$.
- Find the biggest unit fraction different from $\frac{1}{2}$ that is less than or equal to $\frac{5}{14}$. This is $\frac{1}{3}$.
- Find the difference: $\frac{5}{14} - \frac{1}{3} = \frac{1}{42}$.
- Find the biggest unit fraction different from $\frac{1}{2}$ and $\frac{1}{3}$ that is less than or equal to $\frac{1}{42}$. This is $\frac{1}{42}$ itself, so the algorithm stops.

Putting the pieces together, we have

$$\frac{6}{7} = \frac{1}{2} + \frac{1}{3} + \frac{1}{42}$$

which is the required Egyptian representation.

The greedy algorithm doesn't always give the *simplest* Egyptian representation. For instance, when applied to $\frac{5}{121}$ it produces

$$\frac{5}{121} = \frac{1}{25} + \frac{1}{757} + \frac{1}{763,309} + \frac{1}{873,960,180,913} + \frac{1}{1,527,612,795,642,093,418,846,225}$$

and fails to spot a simpler answer:

$$\frac{5}{121} = \frac{1}{33} + \frac{1}{121} + \frac{1}{363}$$

The Erdős–Straus conjecture states that every fraction of the form $4/n$ has a unit fraction representation with three terms:

$$\frac{4}{n} = \frac{1}{x} + \frac{1}{y} + \frac{1}{z}$$

It is true for all $n < 10^{14}$. Exceptions, if they exist, must be very thin on the ground, but no proof or disproof exists.

There are also some interesting variations on the greedy algorithm that you can try. I suggest using fractions with small numerators and denominators to avoid monsters like the one we've just seen. First, try it with the extra condition that every fraction involved must be one over an *even* number. Surprisingly, the greedy algorithm still works – it has been proved that every fraction less than one is a sum of unit fractions with distinct even denominators.

Now try it with *odd* denominators. Computer experiments suggest that it also works in this case. For instance,

$$\frac{4}{23} = \frac{1}{7} + \frac{1}{33} + \frac{1}{1,329} + \frac{1}{2,553,659}$$

But now nobody has a proof. For all we know there might be some peculiar fraction for which the odd-denominator greedy algorithm goes on for ever.

Now that's *really* greedy.

We have only scratched the surface of the mathematics of Egyptian fractions. For more, see:

en.wikipedia.org/wiki/Egyptian_fraction

How to Move a Table

William Feller.

William Feller was a probability theorist at Princeton University. One day he and his wife wanted to move a large table from one room of their house to another, but, try as they might, they couldn't get it through the door. They pushed and pulled and tipped the table on its side and generally tried everything they could, but it just wouldn't go.

Eventually, Feller went back to his desk and worked out a mathematical proof that the table would never be able to pass through the door.

While he was doing this, his wife got the table through the door.

• •

Rectangling the Square

Form five rectangles by choosing their sides from the list 1, 2, 3, 4, 5, 6, 7, 8, 9, 10, with each number being chosen exactly once. Then assemble the rectangles, without overlaps, to form an 11 × 11 square.

Answer on page 293

• •

Newton, by Byron

When Newton saw an apple fall, he found
A mode of proving that the earth turn'd round
In a most natural whirl, called gravitation;
And this is the sole mortal who could grapple
Since Adam, with a fall or with an apple.

Isaac Newton

George Gordon Byron.

X Marks the Spot

'Shiver me timbers and brace the mainsplice!' declared Roger
Redbeard, the pirate captain. 'What have we here, me
hearties? Methinks it be a treasure map, aaargh, for plainly
I sees an X.'

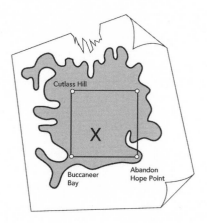

Here be treasure me
hearties, aaargh!

'I knows that island,' said the bosun. 'It be where we marooned that lily-livered swine Admiral Ponsonby-Ffynche and his crew, when we seized the *Vainglorious*. Dead Man's Key, it were. Not a drop o' water on the island, may their bones whiten in the merciless sun.'

'Set sail for Dead Man's Key!' ordered Redbeard. As the crew hoisted the mains'l, he looked round to make sure no one was watching, and turned the map over. On the back, in letters of blood, were instructions to locate the loot:

> Four stone markers form a great square, 140 nautical perches to a side.
> From the markers at Abandon Hope Point, Buccaneer Bay and Cutlass Hill, measure exact whole numbers of nautical perches to the spot marked X.
> From Abandon Ho—
> From Buccaneer Bay: 99 naut—
> From the marker nearest the treasure, Cutl—

The rest was torn away.

Roger cursed a foul pirate curse, for he was a foul pirate and knew how to curse the sort of curse that foul pirates curse. 'I swear,' he swore, 'I'll dig up the entire island if I have to, aaargh!' For he knew that pirates never placed their X's on maps in the correct location, as that would make it too easy for others to discover their hoard.

'If only I'd paid more attention to my maths teacher at school,' Roger sighed. 'For then, by Beelzebub's flaming breeches, I'd know how far X must be from the markers.'*

What are the three distances?

[Hint: This is hard. You may find it helpful to know that if 7 divides a sum of two integer squares, $u^2 + v^2$, then 7 divides each of u and v. Then again ...]

Answer on page 294

• •

Whatever's the Antimatter?

Harold P. Furth was an Austrian-born American physicist who worked on nuclear fusion and related topics. In 2001, he wrote a short poem, 'Perils of Modern Living', which begins:

> Well up above the tropostrata
> There is a region stark and stellar
> Where, on a streak of anti-matter
> Lived Dr Edward Anti-Teller.

Edward Teller was a co-inventor of the hydrogen bomb, acquired huge political influence, and was the inspiration behind the Dr Strangelove character in the movie of the same name. The poem goes on to relate that one day a visitor from Earth turned up, and human and antihuman approached each other:

> ... their right hands
> Clasped, and the rest was gamma rays.

Anyone brought up on *Star Trek* is aware that antimatter is a kind of 'mirror image' of ordinary matter, and when the two are brought into contact they annihilate each other in a gigantic burst of photons ('gamma rays'), particles of light. The combined

---

* He would also have realised that he need dig only along an arc of a circle, centre Buccaneer Bay and radius 99 nautical perches.

mass of the two types of matter is released as energy. Thanks to Einstein's famous formula $E = mc^2$, a small mass m turns into a huge amount of energy E, because the speed of light c is very big, so c^2 is even bigger.

Laying hands on ordinary matter is no great problem; there's a lot of it about. If we could also acquire (*not* by laying hands on) even a small amount of antimatter, we would have a compact source of almost unbounded energy. This potential has long been apparent to physicists and *Star Trek* writers. You just have to find or make antimatter, and store it in something where it won't come into contact with ordinary matter, like a magnetic bottle. It works fine in *Star Trek*, but today's technology falls woefully short of what will be available to starship captains in the 22nd century.*

In the current theories of particle physics, very well supported by experiment, every type of charged subatomic particle has an associated antiparticle, with the same mass but opposite electrical charge, and if the two ever meet … *bang!* Now, *Hoard* is not about physics, but this particular bit of physics came about as an unintended side effect of a mathematical

---

* Or earlier. The warp drive was invented in 2063 by Zefram Cochrane of Alpha Centauri, but the early version used a fusion plasma as an energy source. By the 22nd century and the first series of *Star Trek*, the warp drive was powered by a gravimetric field displacement manifold (or warp core) that used antimatter to create energy. In 1994, in our own universe, the physicist Miguel Alcubierre discovered a 'warp drive' that does not conflict with relativity, yet allows faster-than-light travel. The trick is the oft-repeated science fiction mantra that 'while there is a limit to the speed with which matter can travel through space, there is no limit to the speed with which *space* can travel through space'. Alcubierre found a solution of Einstein's equations for gravity in which the space ahead of a spacecraft contracts while the space behind expands. The spaceship surfs this wave, carried along by a warp bubble of entirely normal space, relative to which it is stationary. Unfortunately, it takes a lot of negative-energy matter to *build* an Alcubierre drive, and we don't have any.

calculation. Sometimes a little bit of maths, taken seriously, can jump-start a scientific revolution.

In 1928, a young physicist named Paul Dirac was trying to reconcile the newfangled ideas of quantum mechanics with the slightly less newfangled ideas of relativity. He focused on the electron, one of the particles out of which atoms are made, and eventually wrote down an equation that both described the quantum properties of this particle and was also consistent with Einstein's special theory of relativity. This, it must be added, was far from easy. The Dirac equation was a major event in physics, and it was one of the discoveries that led to his Nobel Prize in 1933. For all you equation-lovers out there: you'll find it in the note on page 296.

Dirac started from the standard quantum-mechanical equation for the electron, which represents it as a wave; the difficulty was to tinker with this equation so that it respected the requirements of special relativity. To do so, he followed his celebrated nose for mathematical beauty, seeking an equation that treated energy and momentum on the same footing. One evening, sitting beside the fire in Cambridge and musing on this problem, he thought of a clever way to rewrite the 'wave operator' – a key feature of the traditional equation – as the square of something simpler. This step quickly led to some technical issues that were very familiar, and soon the desired equation was staring him in the face.

There was one snag, though. His reformulation introduced new solutions of his equation that did not solve the original version. This always happens when you square an equation; for instance, $x = 2$ becomes $x^2 = 4$ when you square it, and now there is another solution, $x = -2$. Physically, one solution of Dirac's equation has positive kinetic energy,* while the other has negative kinetic energy. The first type of solution obeys all

---

* This is the kind of energy that things acquire when they move, and in classical mechanics it is half the mass times the square of the speed.

requirements for an electron – but what of the second type? On the face of it, negative kinetic energy makes no sense.

In classical (that is, non-quantum) relativity, this kind of thing also happens, but it can be evaded. A particle can never move from a state with positive energy to one with negative energy, because the system must change continuously. So the negative-energy states can be ruled out. But in quantum theory, particles can 'jump' discontinuously from one state to a completely different one. So the electron might, in principle, jump from a physically sensible positive-energy state to one of those baffling negative-energy states.

Dirac decided that he had to allow these puzzling solutions as well. But what were they?

The electron, like all subatomic particles, is characterised by various physical quantities, such as its mass, spin and electrical charge. The particle described by the Dirac equation has all the right properties for an electron; in particular, its spin is $\frac{1}{2}$ and its charge is -1, in suitable units. Working through the details, Dirac noticed that the puzzling solutions were just like electrons, with the same spin and the same mass, but their charge was $+1$, the exact opposite. Dirac had followed his mathematical nose and in effect had predicted a new particle.

Ironically, he stopped short of doing that, partly because he thought the 'new' particle was the familiar proton, which has positive charge. Now, a proton is 1,860 times as heavy as an electron, whereas the negative-energy solution of Dirac's equation has to have the same mass as an electron. But Dirac thought that this discrepancy was caused by some asymmetry in electromagnetism, so he titled his paper 'A Theory of Electrons and Protons'. It was a missed opportunity, because in 1932 Carl D. Anderson spotted a particle with the mass of the electron, but with opposite charge, in an experiment using a cloud chamber to detect cosmic rays. He named the newcomer the *positron*. When asked why he had not predicted the existence of this new particle, Dirac reportedly replied: 'Pure cowardice!'

Not all the difficulties disappeared with the discovery of

positrons. Individual positrons don't have negative kinetic energy, so Dirac suggested that his equation really applies to a 'sea' of negative-energy electrons, which occupy almost all the available negative-energy states. 'An unoccupied negative-energy state,' he wrote, 'will now appear as something with a positive energy, since to make it disappear ... we should have to add to it an electron with negative energy.' And he added that a quantum-mechanical vacuum provides just such a sea of particles. None of this is entirely satisfactory, even when reworked in terms of quantum field theory. But Dirac's equation applies only to a single isolated particle, so it does not describe interactions, which is where the physical discrepancies arise. So physicists are happy to accept the Dirac equation provided that its interpretation is suitably restricted.

The consequences of these discoveries are enormous. Today, particle physicists see the existence of antimatter as a deep and beautiful symmetry in the fundamental laws of nature, called charge conjugation. To every particle there corresponds an antiparticle, differing mainly by having the opposite charge. An uncharged particle, such as the photon, can be its own antiparticle.* If a particle and its antiparticle collide, they annihilate each other in a burst of photons.

The Big Bang ought to have created equal numbers of particles and antiparticles, so our universe ought to contain equal quantities of each type of matter – not counting photons. If the matter and antimatter were thoroughly mixed up, they would collide, so only photons would now exist. However, our universe isn't like that; a lot of matter isn't photons, and all of it seems to be ordinary matter. This is a big puzzle, called baryon asymmetry. No really satisfactory answer to this dilemma has been found. However, it turns out that charge conjugation

---

* However, uncharged particles are not always the same as their antiparticles. The neutron, an uncharged particle, is made from quarks, which individually have non-zero charges. The antineutron is made from the corresponding antiquarks, so the neutron and antineutron are different.

symmetry is not quite exact, and it would have taken only a billion and one particles of matter for every billion particles of antimatter to lead to what we see today. Alternatively, there may be other regions of the universe where antimatter dominates, although that looks rather unlikely. Or maybe time travellers from the distant future may have stolen one particle of antimatter from every billion and one in the early universe, to power their time machines.

Antimatter certainly *exists*, however, because we can make it. Atoms of antihydrogen, made from one positron circling an antiproton, were first created in 1995 at the CERN particle accelerator facility in Geneva. No heavier antiatoms have yet been produced, although the nucleus of antideuterium (an atom that lacks its orbiting positron) has been made. The most common form of antimatter in laboratory experiments is the positron, which can be generated by certain radioactive atoms that undergo beta-plus decay. Here a proton turns into a neutron, a positron and a neutrino. These atoms include carbon-11, potassium-40, nitrogen-13, and others.

The entire Furth poem can be found at:
www.cs.rice.edu/~ssiyer/minstrels/poems/795.html

For more on antimatter physics, see:
en.wikipedia.org/wiki/Antimatter
livefromcern.web.cern.ch/livefromcern/antimatter

For the Alcubierre drive and related topics, see:
en.wikipedia.org/wiki/Alcubierre_drive
hyperspace.wikia.com/wiki/Alcubierre_drive

●●

How to See Inside Things

Antimatter isn't just highbrow physics. Positrons have an important use in medical PET (positron emission tomography) scanners. These are often used in combination with CAT (computerised axial tomography) scanners, often now shortened to CT. Both are based on mathematical techniques invented long ago for no particular practical purpose. Those ideas have to be improved and tweaked, of course, to account for various practical issues – for example, keeping the patient's exposure to X-rays as low as possible, which reduces the amount of data that can be collected.

No,
not like that.

The technology goes back to the early days of X-rays; the mathematics goes back to Johann Radon, who was born in 1887 in Bohemia, which was then part of Austro-Hungary and is now in the Czech Republic. Among his discoveries was the Radon transform.

Johann Radon in 1920.

How to transform him.

The raw material for the Radon transform is a 'function' f defined on all points x of the plane. This means that f defines some rule, which, for any given choice of x, leads to a specific number $f(x)$. Examples are things like 'form the square of x', in which case $f(x) = x^2$, and so on. The transform turns f into a related function F defined on *lines* in the plane. The value $F(L)$ of F at some line L can be thought of as the average of $f(x)$ as x runs along the line.

That's not terribly intuitive (except to professionals), so I'll restate it in terms of something that, in this computer age, may be more familiar. Consider a 'black and white' picture such as the photo of Radon on the opposite page. We can can associate a number with each shade of grey in the picture. So, if $0 =$ white and $1 =$ black, then $\frac{1}{2}$ would be whatever grey you get by mixing equal amounts of black with white, and so on. These numbers determine a 'grey scale': the bigger the number, the darker the shade of grey. So points in Radon's collar are at 0, most of his face is around 0.25 or so, his jacket is 0.5 or higher, and some of the shadows are close to 1.

We can associate a function f with the photo. To do so, let x be any point in the photo, and let $f(x)$ be the number for the shade of grey at that point. For instance, f(point in collar) $= 0$, f(point in face) $= 0.25$, and so on. This function is defined at all points in the plane (within the edges of the photo). We can also reconstruct the photo from the function – in fact, that's how the image is stored in a computer, give or take a few technicalities.

To define the Radon transform F, take any line in the plane – say the line marked L in the right-hand picture. Let $F(L)$ be the average grey-scale value of the photo along line L. Here L cuts across Radon's face, and the average is (say) 0.38. So $F(L) = 0.38$. The line M has a lot more dark grey along it, so maybe $F(M) = 0.72$. You have to do this for every possible line, not just these two: there's a formula for the answer in terms of an integral.

Starting with a function and working out its Radon transform is straightforward, though a bit messy. It is less clear that, given the Radon transform, you can work out the function. Radon's key discovery is that this is possible, and he gave another formula for that calculation. It implies that, if all we know is the average grey-scale value along every line across Radon's photo, then we can work out what Radon looks like.

What does any of this have to do with CAT scans?

Suppose a doctor could take a 'slice' of your body, along a plane, and make a grey-scale image of the tissues that the slice cuts. Dense organs would show up as dark grey, less dense ones as light grey, and so on. It would be just like a plane slice through a sort of 'three-dimensional X-ray' image. And it would tell the doctor exactly where your bodily tissues are, relative to that slice.

Unfortunately, no X-ray machine exists that can take that sort of picture *directly*. But what you can do is pass an X-ray beam – essentially, a straight line – through the body, and measure how strong the radiation is when it comes out at the far side. This strength is related to the average density of tissue – the average grey-scale value of the hypothetical slice – observed along that line. The greater the average tissue density, the weaker the emerging rays are. So, if you shone such a beam along every possible line in the slice plane, you would be able to work out the Radon transform of the grey-scale function for that slice. Then Radon's formula would tell you the grey-scale function itself, and that would be a direct representation of the image created by the plane slice. That is, what that slice of you looks like in real space. So it's a way to see inside solid objects.

In practice you can't measure the Radon transform along *every* line, but you can measure it along enough lines to reconstruct a useful approximation to the image. (Many of the tweaks are to do with this loss of precision.) And this, give or take a few million dollars' worth of technicalities,* is what a CAT scanner does. You lie inside a machine that takes X-ray images from a series of closely spaced angles in a plane that slices

---

* The CAT scanner was pioneered by EMI, primarily a recording company. It is suspected that the millions of dollars came from the sale of Beatles records.

through your body. A computer uses tweaked versions of Radon's formula, or related methods, to work out the corresponding cross-sectional image. The scanner does one more thing: it shifts you along a millimetre or so, and repeats the same process on a parallel slice. And then another, and then another ... building up a full three-dimensional image of your body.

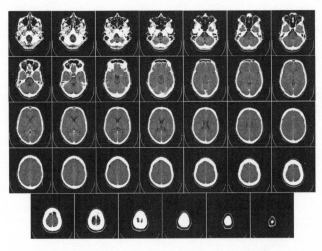

Slices through a human head, made by a CAT scanner.

PET scans use similar technology, and are often performed using the same machines, but with positrons in place of X-rays. The patient is given a dose of a mildly radioactive version of a common body sugar, usually one called fluorodeoxyglucose. The sugar concentrates at different levels in different tissues. As the radioactive element decays, it emits positrons, and the more sugar there is in any location, the more positrons that region emits. The scanner picks up the positrons and measures how much activity there is along any given line. The rest is much as before.

If you ever need a medical scan, it could be worth bearing in mind that what makes it possible is some equations doodled by a mathematical physicist, and a formula discovered nearly a hundred years ago by a pure mathematician interested in a technical question about integral transforms.

•••

Mathematicians Musing About Mathematics

Mathematics is written for mathematicians.   *Nicolaus Copernicus*

Mathematics is the supreme arbiter. From its decisions there is no appeal. We cannot change the rules of the game; we cannot ascertain whether the game is fair.   *Tobias Dantzig*

With me, everything turns into mathematics.   *René Descartes*

Mathematics may be likened to a large rock whose interior composition we wish to examine. The older mathematicians appear as persevering stone-cutters slowly attempting to demolish the rock from the outside with hammer and chisel. The later mathematicians resemble expert miners who seek vulnerable veins, drill into these strategic places, and then blast the rock apart with well placed internal charges.   *Howard W. Eves*

Nature's great book is written in mathematical symbols. *Galileo Galilei*

Mathematics is the queen of the sciences.   *Carl Friedrich Gauss*

Mathematics is a language.   *Josiah Willard Gibbs*

Mathematics is an interesting intellectual sport but it should not be allowed to stand in the way of obtaining sensible information about physical processes.   *Richard W. Hamming*

Pure mathematics is on the whole distinctly more useful than applied. For what is useful above all is technique, and mathematical technique is taught mainly through pure mathematics.   *Godfrey Harold Hardy*

One of the big misapprehensions about mathematics that we perpetrate in our classrooms is that the teacher always seems to know the answer to any problem that is discussed. This gives students the idea that there is a book somewhere with all the right answers to all the interesting questions, and that teachers know those answers. And if one could get hold of the book, one would have everything settled. That's so unlike the true nature of mathematics.   *Leon Henkin*

Mathematics is a game played according to certain simple rules with meaningless marks on paper.   *David Hilbert*

Mathematics is the science of what is clear by itself.
Carl Gustav Jacob Jacobi

Mathematics is the science which uses easy words for hard ideas.   *Edward Kasner and James Newman*

The chief aim of all investigations of the external world should be to discover the rational order and harmony which has been imposed on it by God and which He revealed to us in the language of mathematics.   *Johannes Kepler*

In mathematics you don't understand things. You just get used to them.   *John von Neumann*

Mathematics is the science which draws necessary conclusions.
Benjamin Peirce

Mathematics is the art of giving the same name to different things.   *Henri Poincaré*

We often hear that mathematics consists mainly of 'proving theorems'. Is a writer's job mainly that of 'writing sentences'?
Gian-Carlo Rota

Mathematics may be defined as the subject in which we never know what we are talking about, nor whether what we are saying is true.   *Bertrand Russell*

Mathematics is the science of significant form.
Lynn Arthur Steen

Mathematics is not a book confined within a cover and bound between brazen clasps, whose contents it needs only patience to ransack; it is not a mine, whose treasures may take long to reduce into possession, but which fill only a limited number of veins and lodes; it is not a soil, whose fertility can be exhausted by the yield of successive harvests; it is not a continent or an ocean, whose area can be mapped out and its contour defined: it is limitless as that space which it finds too narrow for its aspirations; its possibilities are as infinite as the worlds which are forever crowding in and multiplying upon the astronomer's gaze.   *James Joseph Sylvester*

Mathematics transfigures the fortuitous concourse of atoms into the tracery of the finger of God.   *Herbert Westren Turnbull*

In many cases, mathematics is an escape from reality. The mathematician finds his own monastic niche and happiness in pursuits that are disconnected from external affairs.
Stanislaw Ulam

God exists since mathematics is consistent, and the Devil exists since we cannot prove it.   *Andre Weil*

Mathematics as a science commenced when first someone, probably a Greek, proved propositions about 'any' things or about 'some' things, without specifications of definite particular things.   *Alfred North Whitehead*

Philosophy is a game with objectives and no rules. Mathematics is a game with rules and no objectives.   *Anonymous*

● ●

Wittgenstein's Sheep

This story is told by the Cambridge analyst John Edensor Littlewood in his lovely little book *A Mathematician's Miscellany*:

> Schoolmaster: 'Suppose *x* is the number of sheep in the problem.'

Schoolboy: 'But, Sir, suppose x is not the number of sheep.'

Littlewood says that he asked the Cambridge philosopher Ludwig Wittgenstein whether this was a profound philosophical joke, and he said it was.

Leaning Tower of Pizza

It was early afternoon in Geronimo's Pizzeria, and business was slow. Angelina, one of the servers, was amusing herself by piling pizza delivery boxes on top of each other on the edge of a table. It all looked rather precarious, and Luigi said as much.

'I'm trying to see how far out I can make the pile go without the boxes actually falling off,' Angelina explained. 'I've discovered that with just three boxes, I can almost get the top one outside the line of the table.'

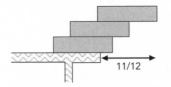

If the boxes are 1 unit long, the top one pokes out 11/12 of a unit.

'How did you figure that out?' Luigi asked.

'Well, I put the top one on the second one, so that its centre was poised exactly on the edge. So it poked out $\frac{1}{2}$ a unit. Then it was obvious that the centre of mass of the top two boxes was in the middle, so I placed them with the centre of mass exactly over the edge of the third box. If you do the sums, that makes it poke out another $\frac{1}{4}$ of a unit. Then I placed the three of them so that their combined centre of mass was right on the edge of the table, and that turned out to add a further $\frac{1}{6}$ of a unit to the overhang.'

'And $\frac{1}{2} + \frac{1}{4} + \frac{1}{6} = \frac{11}{12}$,' Luigi said. 'You're right, it does poke out almost 1 unit.'

Alert readers will observe that Angelina and Luigi are assuming the boxes are identical and they are uniform, that is, the mass is evenly distributed. Real pizza boxes, full or empty, are not like that, but for this puzzle you should pretend that they are.

'What happens if you add more boxes?' Luigi asked.

'I think the pattern continues. I could replace the table by a fourth box, and then slide the pile out until it is just about to topple, adding a further $\frac{1}{8}$ to the overhang. Then the top box *does* poke out over the edge of the table: the overhang is $\frac{25}{24}$. And with more boxes still, I could do the same again, adding $\frac{1}{10}$, and so on.'

'So you're saying,' said Luigi, 'that with n boxes you can get an overhang of

$$\frac{1}{2} + \frac{1}{4} + \frac{1}{6} + \frac{1}{8} + \cdots + \frac{1}{2n}$$

units. Which I instantly recognise as $\frac{1}{2}H_n$, where H_n is the nth harmonic number:

$$H_n = 1 + \frac{1}{2} + \frac{1}{3} + \frac{1}{4} + \cdots + \frac{1}{n}$$

Isn't that right?'

Angelina agreed that it was. As you do.

This is a time-honoured puzzle, and the biggest overhang you can get with n boxes using this method is indeed $\frac{1}{2}H_n$, so Angelina and Luigi are right. You can find the details nicely worked out in many sources, and I would have included them, but for one thing: this traditional answer is valid only with the extra assumption that exactly one box occurs at each level. And that raises a very interesting question: what happens without that assumption?

In 1955, R. Sutton noticed that, even with just three boxes, you can do better than Angelina: an overhang of 1 instead of $\frac{11}{12}$. With four boxes, the biggest possible overhang is

$$\frac{15 - 4\sqrt{2}}{8} = 1.16789$$

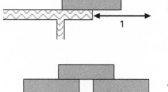

Sutton discovered how to make the top one poke out 1 unit with three boxes

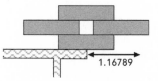

With four boxes, the biggest overlap involves leaving a gap in the second layer.

What happens for n boxes, with as many on each layer as you wish? (There is an even more general question, where the boxes can be tilted, but let's restrict ourselves to layers, like the courses of a brick wall.)

You might like to try your hand at this puzzle before reading any further. What is the biggest overhang you can get with 5 or 6 boxes?

Answers on page 297

To avoid misunderstandings, let me make the conditions clear. All boxes are identical and uniform, and everything is idealised to exact rectangles and all the usual stuff we assume in Euclidean geometry. The problem is posed in the plane, because in three-dimensional space you could also rotate boxes, without violating the 'layers' condition. The arrangement must be in equilibrium: that is, if you work out all the forces that act on any box, they all balance each other out. Boxes must be arranged in layers, but you can leave gaps. And one other important condition: you do not have to be able to build the arrangement by adding one box at a time. Intermediate stages might topple if left unsupported. Only the final arrangement must be in equilibrium. (This equilibrium condition turns out not to be terribly intuitive; it can be turned into equations and checked by computer. When there aren't too many boxes, though, it should be intuitive enough for you to tackle this puzzle.)

The answers for 4, 5 and 6 boxes were worked out by J. F. Hall in 2005. In fact, he proposed some general patterns, and suggested that they should always maximise the overhang. But, in 2009, Mike Paterson and Uri Zwick showed that Hall's stacks maximise the overhang only for 19 boxes or fewer (see page 297 for the reference). Finding exact arrangements with a lot of boxes is extremely complicated, but they proposed some near-optimal arrangements for up to 100 boxes.

One very interesting question is: how fast can the biggest overhang grow as the number n of boxes increases? For the classic 'one box per layer' solution, the answer is $\frac{1}{2}H_n$. There doesn't seem to be a simple formula for this number, but H_n is very closely approximated by the natural logarithm $\log n$. So the 'asymptotic' size of the largest overhang is $\frac{1}{2}\log n$.

Paterson and Zwick proved that, when layers can contain many boxes, the maximal overhang is approximately proportional to the cube root of n. More precisely, there are constants c and C for which the maximal overhang always lies between $c\sqrt[3]{n}$ and $C\sqrt[3]{n}$. They exhibited explicit arrangements with an overhang of at least

$$\sqrt[3]{\frac{3}{16}}\sqrt[3]{n} - \frac{1}{4} = 0.572357\sqrt[3]{n} - \frac{1}{4}$$

units, using what they call 'parabolic stacks'. The picture shows such a stack with 111 boxes and an overhang of exactly 3 units. (The approximate formula gives only 2.50069 instead of 3 when $n = 111$, but it still gives the best-known overhang for very large n.)

Early in 2009, Peter Winkler, Yuval Peres and Mikkel Thorup joined the team, and took the question further. They proved that C is at most 6: the overhang can never be greater than $6\sqrt[3]{n}$. Their proof uses the probability theory of 'random walks', in which a person takes a step forward or backward with specified probabilities. Each new brick spreads the forces that act in a similar manner to the way probabilities spread as a random walk proceeds.

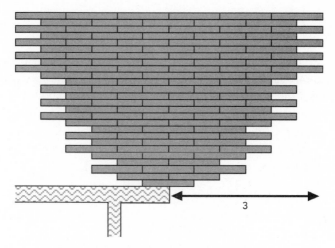

A parabolic stack with 111 boxes, and overhang 3.

• •

PieThagoras's World-Famous Mince πs

Alvin, Brenda and Casimir went to the pie-shop and bought three of PieThagoras's world-famous perfectly circular mince pies. They bought one mini-pie with diameter 6 centimetres, one midi-pie with diameter 8 centimetres, and one maxi-pie with diameter 10 centimetres, because those were the only pies left.

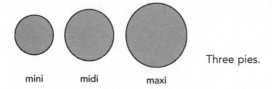

Three pies.

mini   midi   maxi

They could have settled for one pie each, but they wanted to share the pies fairly. Now, as everyone knows, PieThagoras's world-famous mince pies consist of two flat layers of pastry, of uniform thickness, with a uniform layer of mince sandwiched in between. The thicknesses of the pastry and the mince are the same for all sizes of pie. So 'fair' means 'having equal area' when viewed from above as in my picture.

They decided that sharing the pies fairly would be quite complicated, and had just settled on dividing each pie separately into thirds when Desdemona turned up and demanded her fair share too. Fortunately they had not started cutting the pies. After some thought, they discovered that now they could divide the pies more easily, by cutting two of them into two pieces each, and leaving the third pie uncut. How?

Answer on page 297

. . . .

Diamond Frame

Innumeratus's attempt at a magic frame.

Innumeratus had taken the ace to 10 of diamonds from a pack of cards, and was arranging them to make a rectangular frame.

'Look!' he shouted to Mathophila. 'I've arranged them so that the total number of pips along each side of the frame is the same!'

Mathophila had learned to take such statements with a pinch of salt, and she quickly pointed out that the sums concerned were 19 (top), 20 (left), 22 (right) and 16 (bottom).

'Well, I've arranged them so that the total number of pips along each side of the frame is *different*, then.'

Mathophila agreed with that, but felt it was a silly puzzle. She really liked the first version better.

Can you solve the original version? You can turn cards through a right angle if you wish.

Answer on page 298

Pour Relations

This is a traditional puzzle that goes back to the Renaissance Italian mathematician Tartaglia in the 1500s, but its solution has systematic features that were not noticed until 1939, discussed in the Answer. There are many similar puzzles.

You have three jugs, which respectively hold 3 litres, 5 litres, and 8 litres of water. The 8-litre jug is full, the other two are empty. Your task is to divide the water into two parts, each of 4 litres, by pouring water from one jug into another. You are not allowed to estimate quantities by eye, so you can only stop pouring when one of the jugs involved becomes either full or empty.

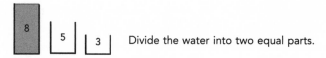

8   5   3   Divide the water into two equal parts.

Answer on page 299

Alexander's Horned Sphere

If you draw a closed curve in the plane which doesn't cross itself, then it seems pretty obvious that it must divide the plane into two regions: one inside the curve, the other outside it. But mathematical curves can be very wiggly, and it turns out that this obvious statement is difficult to prove. Camille Jordan gave an attempted proof, more than 80 pages long, in a textbook published in several volumes between 1882 and 1887, but it turned out to be incomplete. Oswald Veblen found the first correct proof of this 'Jordan curve theorem' in 1905. In 2005, a team of mathematicians developed a proof suitable for computer verification – and verified it. The proof was 6,500 lines long.

A closed curve, with the inside shaded.

A subtler topological feature of such a closed curve is that the regions inside and outside the curve are topologically equivalent to the regions inside and outside an ordinary circle. This too may seem obvious, but, remarkably, the corresponding statement in three dimensions, which seems equally obvious, is actually false. That is: there is a surface in space, topologically equivalent to an ordinary sphere, whose inside is topologically equivalent to the inside of an ordinary sphere, but whose outside is *not* topologically equivalent to the outside of an ordinary sphere! Such a surface was discovered by James Waddell Alexander in 1924, and is called *Alexander's horned sphere*. It is like a sphere that has sprouted a pair of horns, which divide repeatedly and intertwine.

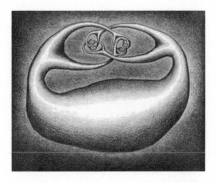

Alexander's horned sphere.

The Sacred Principle of Mat

The daredevil adventurer and treasure-hunter Colorado Smith, who is not like a real archaeologist *at all*, ducked a passing shower of blazing war-arrows to check the crude sketch-map scribbled in his father's battered notebook.

'The holy sanctuary of Pheedme-Pheedme the goddess of eating and sleeping,' he read, 'is formed from 64 identical square cushions stuffed with ostrich-down, arranged in an 8 × 8 array. The five sacred avatars of Pheedme-Pheedme, represented in overstuffed fabric, are to be placed on the cushions so that they "watch over" every other cushion: that is, every other cushion must be in line with one occupied by an avatar. This line can be horizontal, vertical or diagonal, relative to the array, where "diagonal" means "sloping at 45° ".'

'Look out!' shrieked his sidekick Brunnhilde, taking cover beneath a large stone altar.

'I wouldn't do that if I were you,' said Smith, and hauled her out a split second before the supporting slabs exploded in puffs of dust and the 10-tonne altar stone crashed to the ground. 'Now, Dad's notebook says something about the principles of – uh – *Mat*?'

'Ma'at was the Egyptian concept of justice and rightful place,' Brunnhilde pointed out. 'But this temple is Burmalayan.'

'True. Can't be ma'at ... No, it's definitely the Principle of Mat. Apparently the goddess reclines on a mat, surrounded by her sacred avatars. We have to leave a space for the holy reclining mat, which is square. Hmm ... maybe *this* would do.'

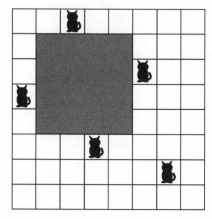

Is this how to lay out the five sacred avatars and Pheedme-Pheedme's reclining mat?

'That seems suspiciously easy,' said Brunnhilde. 'What *else* do we have to do?'

Smith quietly removed a deadly kamikaze-scorpion from her hair, hoping she wouldn't notice. 'Uh, we have to arrange the avatars to leave the largest possible space for a square mat. Bearing in mind that they must watch over every cushion. I doubt we can do better than my picture.'

'Those ancient priests were sneaky, though,' said Brunnhilde. She tried not to listen to the approaching bloodcurdling cries, and racked her brains. If they could solve the riddle of the sacred mat, they could proceed to the enigma of the potted dormice, and then only 17 more puzzles would lie between them and the Hoard of Treasures. 'Does the mat have to be arranged with its sides parallel to those of the cushions? Could it be *tilted*?'

'I don't see anything in the 999 pages of *The Book of the Ninth Life* to prohibit that,' said Smith. 'The only restriction is that the mat can't overlap any cushion bearing one of the sacred avatars. The edges of the mat and the cushion can touch, but there mustn't be a genuine overlap.'

How can the biggest mat be fitted in without breaking the sacred rules?

Answer on page 301

• •

Perfectly Abundantly Amicably Deficient

If n is a whole number, then the sum of its divisors, including n itself, is the divisor-sum $\sigma(n)$. So, for example,

$$\sigma(24) = 1 + 2 + 3 + 4 + 6 + 8 + 12 + 24 = 60$$

The divisor-sum is the key to a very ancient recreation, the search for perfect numbers. A number is *abundant* if it is smaller than the sum of its 'proper' divisors – those that exclude the number itself. It is *deficient* if it is greater than that sum, and *perfect* if it is equal to it. In terms of the divisor-sum, these conditions become

$$\sigma(v) > 2v \qquad \sigma(v) < 2v \qquad \sigma(v) = 2v$$

Here we see $2n$, rather than n, because $\sigma(n)$ includes the divisor n as well as all the others. This is done so that the nice formula $\sigma(mn) = \sigma(m)\sigma(n)$ holds when m and n have no common factor greater than 1.

Lots of numbers are deficient; for example 10 has proper divisors 1, 2, 5, which sum to 8. Abundant numbers are rarer: 12 has proper divisors 1, 2, 3, 4, 6, with sum 16. Perfect numbers are very rare; the first few are:

$$6 = 1 + 2 + 3$$
$$28 = 1 + 2 + 4 + 7 + 14$$

followed by 496 and 8,128. Euclid discovered a pattern in these perfect numbers: he proved that whenever $2^p - 1$ is a prime, the number $2^{p-1}(2^p - 1)$ is perfect. Much later, Euler proved that every *even* perfect number must be of this form. Primes of the form $2^p - 1$ are called Mersenne primes (*Cabinet*, page 151).

No one knows whether any odd perfect numbers exist;

however, Carl Pomerance has given a non-rigorous but plausible argument that they don't. There is a solid proof that if an odd perfect number does exist then it must be at least 10^{300}, and have at least 75 prime factors. Its largest prime factor must be greater than 10^8.

A related, equally ancient pastime is to find pairs of *amicable* numbers – each equal to the sum of the proper divisors of the other. That is,

$$m = \sigma(v) - v$$
$$n = \sigma(\mu) - \mu$$

so $\sigma(n) = \sigma(m) = m + n$. For example, the proper divisors of 220 are 1, 2, 4, 5, 10, 11, 20, 22, 44, 55, 110, adding to 284; the proper divisors of 284 are 1, 2, 4, 71, 142, adding to 220. The next few amicable pairs are (1184, 1210), (2620, 2924), (5020, 5564) and (6232, 6368).

In all known examples, the numbers in an amicable pair are either both even or both odd. Every known pair shares at least one common factor; it is not known whether a pair of amicable numbers with no common factor can exist. If there is such pair, then their product is at least 10^{67}.

An integer is *multiply perfect* if it divides the sum of its divisors exactly; the *multiplicity* is the quotient. Here it makes no difference whether we include the number itself, or not, except that the multiplicity goes down by 1 if we don't. But it's normal to include it. If we do, then ordinary perfect numbers have multiplicity 2, triperfect numbers have multiplicity 3, and so on. The smallest triperfect number is 120, as Robert Recorde knew in 1557: the sum of its divisors is

$$1 + 2 + 3 + 4 + 5 + 6 + 8 + 10 + 12 + 15 + 20 + 24 + 30$$
$$+ 40 + 60 + 120$$
$$= 360 = 3 \times 120$$

Here are a few other multiply perfect numbers. Many more are known. (The dots between the numbers mean 'multiply'.)

Number	Discoverer	Date
Triperfect		
$2^3.3.5$	Robert Recorde	1557
$2^5.3.7$	Pierre de Fermat	1636
$2^9.3.11.31$	André Jumeau de Sainte-Croix	1638
$2^8.5.7.19.37.73$	Marin Mersenne	1638
Quadruperfect		
$2^5.3^3.5.7$	René Descartes	1638
$2^3.3^2.5.7.13$	René Descartes	1638
$2^9.3^3.5.11.31$	René Descartes	1638
$2^8.3.5.7.19.37.73$	Édouard Lucas	1891
Quintuperfect		
$2^7.3^4.5.7.11^2.17.19$	René Descartes	1638
$2^{10}.3^5.5.7^2.13.19.23.89$	Bernard Frénicle de Bessy	1638
Sextuperfect		
$2^{23}.3^7.5^3.7^4.11^3.13^3.17^2.31.$ $41.61.241.307.467.2801$	Pierre de Fermat	1643
$2^{27}.3^5.5^3.7.11.13^2.19.29.31.$ $41.43.61.113.127$	Pierre de Fermat	1643
Septuperfect		
$2^{46}(2^{47}-1).19^2.127.\ 3^{15}.5^3.7^5.$ $11.13.17.23.31.37.41.43.$ $61.89.97.193.442151$	Allan Cunningham	1902

Target Practice

Robin Hood and Friar Tuck were engaging in some target practice. The target was a series of concentric rings, lying between successive circles with radii 1, 2, 3, 4, 5. (The innermost circle counts as a ring.)

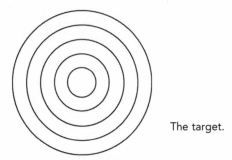

The target.

Friar Tuck and Robin both fired a number of arrows.

'Yours are all closer to the centre than mine,' said Tuck ruefully.

'That's why I'm the leader of this outlaw band,' Robin pointed out.

'But let's look on the bright side,' Tuck replied. 'The total area of the rings that I hit is the same as the total area of the rings you hit. So that makes us equally accurate, right?'

Naturally, Robin pointed out the fallacy ... but:

Which rings did the two archers hit?

(A ring may be hit more than once, but it only counts once towards the area.)

For a bonus point: what is the smallest number of rings for which this question has two or more different answers?

For a further bonus point: if each archer's rings are adjacent – no gaps where a ring that has not been hit lies between two that have – what is the smallest number of rings for which this question has two or more different answers?

Answers on page 302

● ●

Just a Phase I'm Going Through

Over the course of one lunar month, the phases of the Moon run from new moon to full moon and back again, passing through various intermediate shapes known as 'crescent', 'first quarter', 'waning gibbous', and the like.

new Moon   crescent   first quarter   waxing gibbous   full Moon   waning gibbous   third quarter   crescent   new Moon

The two 'quarter' moons are so named because they occur one-quarter and three-quarters of the way through the lunar month, starting from a new moon. At these times the area of the visible part is half the Moon's face, not one-quarter. But there are two times during the cycle where a crescent moon occurs, whose visible area is exactly one-quarter of the area of the lunar disc.

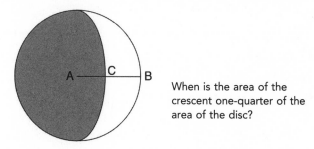

When is the area of the crescent one-quarter of the area of the disc?

- When this happens, what fraction of the radius AB is the width CB of the crescent?
- At which fractions of a full cycle, starting from the new moon, do these special crescents occur?

To simplify the geometry, assume that the Moon is a sphere, and the orbits of both the Moon (round the Earth) and the Earth (round the Sun) are circles lying in the same plane, with both bodies moving at a constant speed. Then the length of a lunar month will also be constant. Assume, too, that the Sun is so far away that its rays are all parallel, and the Moon is sufficiently distant for its image as seen from Earth to be obtained by parallel projection – as if every point on the Moon were transferred to a screen along a line meeting the screen at right angles. (However, you have to replace the real Moon by a much smaller one, otherwise its image 'in' the eye would be 3,474 kilometres, or 2,159 miles, across.)

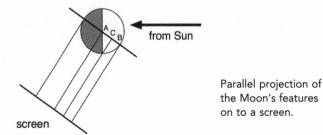

from Sun

Parallel projection of
the Moon's features
on to a screen.

screen

None of these assumptions is true, but they're good
approximations, and the geometry gets a lot harder without
them.

Answers on page 303

Proof Techniques

- *Proof by Contradiction*: 'This theorem contradicts a well-known result due to Isaac Newton.'
- *Proof by Metacontradiction*: 'We prove that a proof exists. To do so, assume that there is no proof...'
- *Proof by Deferral*: 'We'll prove this next week.'
- *Proof by Cyclic Deferral:* 'As we proved last week ...'
- *Proof by Indefinite Deferral*: 'As I said last week, we'll prove this next week.'
- *Proof by Intimidation*: 'As any fool can see, the proof is obviously trivial.'
- *Proof by Deferred Intimidation*: 'As any fool can see, the proof is obviously trivial.' 'Sorry, Professor, are you sure?' Goes away for half an hour. Comes back. 'Yes.'
- *Proof by Handwaving*: 'Self-explanatory.' Most effective in seminars and conference talks.
- *Proof by Vigorous Handwaving*: More tiring, but more effective.
- *Proof by Over-optimistic Citation*: 'As Pythagoras proved, two cubes never add up to a cube.'

- *Proof by Personal Conviction*: 'It is my profound belief that the quaternionic pseudo-Mandelbrot set is locally disconnected.'
- *Proof by Lack of Imagination*: 'I can't think of any reason why it's false, so it must be true.'
- *Proof by Forward Reference*: 'My proof that the quaternionic pseudo-Mandelbrot set is locally disconnected will appear in a forthcoming paper.' Often not as forthcoming as it seemed when the reference was made.
- *Proof by Example*: 'We prove the case $n = 2$ and then let $2 = n$.'
- *Proof by Omission*: 'The other 142 cases are analogous.'
- *Proof by Outsourcing*: 'Details are left to the reader.'
- *Statement by Outsourcing*: 'Formulation of the correct theorem is left to the reader.'
- *Proof by Unreadable Notation*: 'If you work through the next 500 pages of incredibly dense formulas in six alphabets, you'll see why it has to be true.'
- *Proof by Authority*: 'I saw Milnor in the cafeteria and he said he thought it's probably locally disconnected.'
- *Proof by Personal Communication*: 'The quaternionic pseudo-Mandelbrot set is locally disconnected (Milnor, personal communication).'
- *Proof by Vague Authority*: 'The quaternionic pseudo-Mandelbrot set is well known to be locally disconnected.'
- *Proof by Provocative Wager*: 'If the quaternionic pseudo-Mandelbrot set is not locally disconnected, I'll jump off London Bridge wearing a gorilla suit.'
- *Proof by Erudite Allusion*: 'Local connectivity of the quaternionic pseudo-Mandelbrot set follows by adapting the methods of Cheesburger and Fries to non-compact infinite-dimensional quasi-manifolds over skew fields of characteristic greater than 11.'
- *Proof by Reduction to the Wrong Problem*: 'To see that the quaternionic pseudo-Mandelbrot set is locally disconnected, we reduce it to Pythagoras's Theorem.'
- *Proof by Inaccessible Reference*: 'A proof that the quaternionic pseudo-Mandelbrot set is locally disconnected can be easily

derived from Pzkrzwcziewszczii's privately printed memoir bound into volume $1\frac{1}{2}$ of the printer's proofs of the 1831 *Proceedings of the South Liechtenstein Ladies' Knitting Circle* before the entire print run was pulped.'

• •

Second Thoughts

'This is a one-line proof – if we start sufficiently far to the left.'

• •

How Dudeney Cooked Loyd

In *Mathematical Carnival*, the celebrated recreational mathematician Martin Gardner tells us: 'When a puzzle is found to contain a major flaw – when the answer is wrong, when there is no answer, or when, contrary to claims, there is more than one answer or a better answer – the puzzle is said to be "cooked".' Gardner gives several examples, the simplest being a puzzle he had set in a children's book. In the array of numbers

9	9	9
5	5	5
3	3	3
1	1	1

circle six digits to make the total of circled numbers equal 21. See page 304 for Gardner's answer, why he had to cook the puzzle, how he did that, and how one of his readers cooked his cook. Both solutions are what Gardner calls a quibble-cook, because they exploit an imprecise specification in the question.

Gardner, a puzzle expert, also mentions a more serious example of cookery involving the two arch-rivals of late 19th and early 20th century puzzling, the American Sam Loyd and the Englishman Henry Ernest Dudeney. The problem was to cut a mitre (a square with one triangular quarter missing) into as few pieces as possible, so that they could be rearranged to make a perfect square. Loyd's solution was to cut off two small triangles and then use a 'staircase' construction – four pieces in all.

After Loyd had published his solution in his *Cyclopaedia of Puzzles*, Dudeney spotted an error, and found a correct solution with five pieces. The easier question here is: what was the mistake? The harder one is to put it right.

Answers on page 304

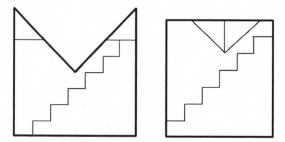

Loyd's four-piece attempt to dissect a mitre into a square.

Cooking with Water

Speaking of quibble-cooks: I'm going to set exactly the same puzzle as one I set in *Cabinet* (page 199), where the answer was 'impossible'. But I'm looking for a different answer, because this time I'll allow any suitably clever quibble-cook.

Three houses have to be connected to three utility companies – water, gas, electricity. Each house must be connected to *all three* utilities. Can you do this without the connections crossing? (Work in the plane, no third direction to pass pipes over or under cables. And you are not allowed to pass the cables through a house or a utility company building.)

Actually, I should have said: 'You are not allowed to pass the cables *or pipes* through a house or a utility company building.' I

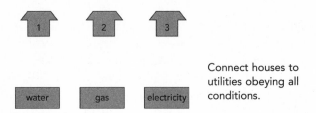

Connect houses to utilities obeying all conditions.

think that was clear from the context, but just in case you disagree, assume that too.

Answer on page 305

• •

Celestial Resonance

In the earliest days of the telescope, Galileo Galilei discovered that the planet Jupiter had four moons, now named Io, Europa, Ganymede and Callisto. Astronomers now know of at least 63 moons of Jupiter, but the rest are much smaller than these four 'Galilean' satellites, and some are very small indeed. The times the Galilean satellites take to go once round Jupiter, in days, are respectively 1.769, 3.551, 7.155 and 16.689. What is remarkable about these numbers is that each is roughly twice the previous one. In fact,

$$3.551/1.769 = 2.007$$
$$7.155/3.551 = 2.015$$
$$16.689/7.155 = 2.332$$

The first two ratios are very close to 2; the third one is less impressive.

The simple numerical relationships between the first three periods are not coincidental: they result from a dynamic *resonance*, in which configurations of moons or planets tend to repeat themselves at regular periods. Europa and Io are in 2:1 resonance, and so are Ganymede and Europa. The ratio is that of the orbital periods of the two moons concerned; the numbers of orbits they make in the same time are in the opposite ratio, 1:2.

Resonances arise because the corresponding orbits are especially stable, so they are not disrupted by any other bodies in the vicinity, such as the other moons of Jupiter. However, to make things more difficult, some types of resonance are especially unstable, depending on the ratio concerned and the physical system involved. We don't fully understand the reasons for this. But this type of 2:1 resonance is very stable, and this is why we find it in Jupiter's larger moons.

The other main orbital resonances within the Solar System are:

- 3:2 Pluto–Neptune — 90,465 and 60,190.5 days
- 2:1 Tethys–Mimas — 1.887 and 0.942 days
- 2:1 Dione–Enceladus — 2.737 and 1.370 days
- 4:3 Hyperion–Titan — 21.277 and 15.945 days

where all bodies listed except Pluto and Neptune are moons of Saturn.*

When thinking about resonances, it is important to realise that *any* ratio can be approximated by exact fractions, and there can be 'accidental resonances' that are unrelated to dynamic influences between the two orbits concerned. All the above are genuine resonances, showing features such as 'precession of perihelion' – movement of the orbital position nearest to the Sun – that lock the orbits stably together. Among the accidental resonances that can be found by searching tables of astronomical data are:

- 13:8 Earth–Venus
- 3:1 Mars–Venus
- 2:1 Mars–Earth
- 12:1 Jupiter–Earth

---

* As of 2006, the International Astronomical Union declared that Pluto is no longer considered a 'planet', but a 'dwarf planet' or 'plutoid'. Not all astronomers approve of this.

- 5 : 2 Saturn–Jupiter
- 7 : 1 Uranus–Jupiter
- 2 : 1 Neptune–Uranus

Some important genuine resonances occur for asteroids – mainly small bodies, most of which orbit between Mars and Jupiter. Resonances with Jupiter cause asteroids to 'clump' at some distances from the Sun, and to avoid other distances.* More asteroids than average have orbits that are in 2 : 3, 3 : 4 and 1 : 1 resonance with Jupiter (the Hilda family, Thule, and the Trojans) because these resonances stabilise the orbits. In contrast, the resonances 1 : 3, 2 : 5, 3 : 7 and 1 : 2 destabilise the orbits: rings and belts are different from individual bodies. As a result, there are very few asteroids at the corresponding distances from the Sun, called Kirkwood gaps.

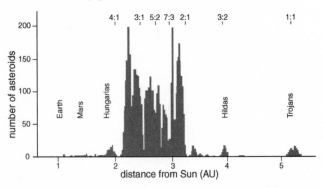

Kirkwood gaps and Hilda clumps (1 AU is the Earth–Sun distance).

Similar effects occur in Saturn's rings. For instance, the Cassini Division – a prominent gap in the rings – is caused by a

---

* 'Clump' is a metaphor: they don't look like the asteroid belts in *Star Wars*, and what clumps is the *distances*, not the asteroids themselves. Actually, nowhere in the asteroid belt looks like an asteroid belt in *Star Wars*. If you stood on a typical asteroid and looked around for the nearest one, it would be about a million miles (1.6 million kilometres) away. No exciting chase scenes, then.

2 : 1 resonance with Mimas, which this time is unstable. The 'A ring' does not slowly fuzz out, because a 6 : 7 resonance with Janus sweeps material away from the outer edge.

One of the weirdest resonances occurs in the rings of Neptune, a ratio of 43 : 42. Despite the big numbers, this one seems to be a genuine dynamic effect. Neptune's Adams ring is a complete, though narrow, ring, and it is much denser in some places than others, so the dense regions create a series of short arcs. The problem is to explain how these arcs are spaced along that orbit, and a 43 : 42 resonance with the moon Galatea, which lies just inside the Adams ring, is thought to be responsible. The arcs should then be placed at some of the 84 equilibrium points associated with this resonance, which form the vertices of a regular 84-sided polygon, and pictures from *Voyager 2* support this.

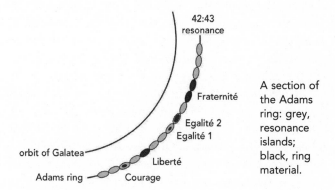

A section of the Adams ring: grey, resonance islands; black, ring material.

Resonances are not confined to the orbital periods of moons and planets. Our own Moon always turns the same face towards our planet, so that the 'far side' remains hidden. The Moon wobbles a bit, but 82 per cent of the far side is never visible from Earth. This is a 1 : 1 resonance between the Moon's period of rotation around its axis and its period of revolution around the Earth. This type of effect is called spin–orbital resonance, and again there are plenty of examples. It used to be thought that the planet Mercury did the same as our Moon, so that one side –

facing the Sun – was tremendously hot, and the other one tremendously cold. This turned out to be a mistake, caused by the difficulty of observing the planet when it was close to the Sun and the absence of any surface markings visible in the telescopes then available. In fact, the periods of revolution and rotation of Mercury are 87.97 days and 58.65 days, with a ratio of 1.4999 – a very precise 3 : 2 resonance.

Astronomers now know that many stars also have planets; indeed, a total of 344 'extrasolar' planets has been found[*] since the first one was detected in 1989. For example, two planets of the star Gliese 876, known as Gliese 876b and Gliese 876c, are in 2 : 1 resonance. Extrasolar planets are normally detected either by their (tiny) gravitational effects on their parent star, or by changes in the star's light as and if the planets pass across its face as seen from Earth. But in 2007 the first telescopic image of such a planet was obtained, around a star rejoicing in the name HR8799.[†] The main difficulty here is that the light from the star swamps that of the planet, so various mathematical techniques are used to 'subtract' the star's light. Early in 2009, it was discovered that one of these stars can be extracted, by similar image-processing methods, from a photo of the star taken by the Hubble telescope in 1998, but that's an aside. The point is that the dynamics of this three-planet system is unstable, so that we would be unlikely to observe it unless the planets are in 4 : 2 : 1 resonance. An important consequence of this line of thinking is that such resonances improve the chances of other stable planetary systems existing. Which, perhaps, also improves the prospect of alien life existing somewhere.

A good site for this topic is:
en.wikipedia.org/wiki/Orbital_resonance

---

[*] As of 1 April 2009. See *The Extrasolar Planets Encyclopaedia* at **www.exoplanet.eu** for the latest information.

[†] That is, it is item HR8799 in the *Yale Bright Star Catalogue*. The HR prefix refers the earlier *Harvard Revised Photometry Catalogue*, and most of the stars listed in the Yale catalogue come from this.

which has a long list of 'accidental' resonances, more explanation of the dynamics involved, and an animation of Jupiter's moons' $1:2:4$ resonance. There is also an animation showing how planets cause a star's position to 'wobble' at: www.gavinrymill.com/dinosaurs/extra-solar-planets.html

• •

Calculator Curiosity 2

What's special about the number 0588235294117647? (That leading zero does matter here.) Try multiplying it by 2, 3, 4, 5, 6, 7, 8, 9, 10, 11, 12, 13, 14, 15, 16, and you'll see. You do need a calculator or software that works with 16-digit numbers. I find that the human brain, a piece of paper and a pen does that job pretty well.

What happens when you multiply it by 17?

Answer on page 305

• •

Which is Bigger?

Which is bigger: e^π or π^e?

They are surprisingly close together. Recall that $e \approx 2.71828$ and $\pi \approx 3.14159$.

Answer on page 306

• •

Sums That Go On For Ever

They sound like a childhood nightmare, but sums where you never get to the end are among the most important mathematical inventions. Of course, you don't work them out by doing an infinitely long calculation, but, conceptually, they open up very powerful practical ways to calculate things that mathematicians and scientists want to know.

Back in the 18th century, mathematicians were coming to grips with – or often *not* coming to grips with – the paradoxical behaviour of infinite sums (or *series*). They were happy to use sums like

$$1 + \frac{1}{2} + \frac{1}{4} + \frac{1}{8} + \frac{1}{16} + \frac{1}{32} + \cdots$$

(where the $\cdots$ means that the series never stops) and they were also happy that this particular sum is exactly equal to 2. Indeed, if the sum is s, then

$$2s = 2 + 1 + \frac{1}{2} + \frac{1}{4} + \frac{1}{8} + \frac{1}{16} + \cdots = 2 + s$$

so $s = 2$.

However, the innocuous-looking series

$$1 - 1 + 1 - 1 + 1 - 1 + \cdots$$

is a different matter. Bracketed like this:

$$(1 - 1) + (1 - 1) + (1 - 1) + \cdots$$

it reduces to $0 + 0 + 0 + \cdots$, which surely must be 0. But bracketed like *this*:

$$1 + (-1 + 1) + (-1 + 1) + (-1 + 1) + \cdots$$

it becomes $1 + 0 + 0 + 0 + \cdots$, which surely must be 1. (The extra $+$ signs in front of the brackets are there because the minus sign does double duty: both as an instruction to subtract, and to denote a negative number.) No less a figure than the great Euler tried the same sort of trick that we used to sum the first series, letting s be the sum and manipulating the series to get an equation for s. He observed that

$$s = 1 - 1 + 1 - 1 + 1 - 1 + \cdots$$
$$= 1 - (1 - 1 + 1 - 1 + 1 - 1 + \cdots) = 1 - s$$

and argued that $s = \frac{1}{2}$.

This is a nice compromise between the conflicting values 0 and 1 that we've just found, but at the time Euler's suggestion just muddied the waters further. And they were already fairly murky. The first satisfactory answer was to distinguish between *convergent* series, which settle down closer and closer to some specific number, and *divergent* ones, which don't. For instance, successive steps in the first series give the numbers

$$1, \quad \frac{3}{2}, \quad \frac{7}{4}, \quad \frac{15}{8}, \quad \frac{31}{16}, \quad \cdots$$

which get ever closer to 2 (and *only* to 2). So this series converges, and its sum is defined to be 2. However, the second series leads to successive sums

$$1, \quad 0, \quad 1, \quad 0, \quad 1, \quad \cdots$$

which hop to and fro, but never settle down near any specific number. So that series is divergent. Divergent series were declared taboo, because they couldn't safely be manipulated using the standard rules of algebra. Convergent series were better behaved, but even those sometimes had to be handled with care.

Much, much later it turned out that there are clever 'summation methods' that can assign a meaningful sum to certain divergent series, in such a way that appropriate versions of the standard rules of algebra still work. The key to these methods lies in the interpretation placed on the series, and I don't want to dig into the rather technical ideas involved, except to record that Euler's controversial $\frac{1}{2}$ can be justified in such a setting. In astronomy, another approach led to a theory of 'asymptotic series' that can be used to calculate positions of planets and so on, even though the series diverge. These ideas proved useful in several other areas of science.

The first message here is that, whenever a traditional concept in mathematics is extended into a new realm, it is worth asking whether the expected features persist, and often the answer is 'some do, some don't'. The second message is: don't give up on a good idea, just because it doesn't work.

The Most Outrageous Proof

The Great Whodunni, with the assistance of Grumpelina, produces a length of soft rope from thin air and ties a knot in it. A little further along, he ties a second knot. Holding the two free ends in each hand, he gives the rope a shake – and the knots disappear.

Mathematically, it's obvious, of course. The second knot must be the anti-knot of the first one. You just tie it so that all the twists and turns cancel out. Right?

Wrong. Topologists know that there is no such thing as an anti-knot.

To be sure, there are very complicated knots that turn out not to be knotted at all. But that's a different issue. What you can't do is tie two genuine (un-unknottable) knots in the same piece of rope, clearly separated from each other, and then deform the whole thing into an unknotted piece of rope. Not if the ends of the rope are glued together or otherwise pinned down so that the knots can't escape.

Not only do topologists know that: they can prove it. The first proofs were really complicated, but eventually someone found a very simple proof. Which is completely outrageous. You probably won't believe it when I show it to you. Especially not when we've just been exposed to the paradoxical properties of infinite series.

A mathematician's knot is a closed curve in space, and it is genuinely knotted if it can't be continuously deformed into a circle – the archetypal *unknotted* closed curve. Real knots are tied in lengths of string that have ends, and the only reason we can tie them at all is because the ends can poke through loops to create the knot. However, the topology of such 'knots' isn't very interesting, because they can all be unknotted. So mathematicians have to redefine knots to stop them falling off the ends of the string. Gluing the ends into a circle is one method, but there's another one: put the knot inside a box and glue the ends to the walls of the box. If the string stays inside the box, the knot can't escape over the ends. (The box can be any size and shape provided it is topologically equivalent to a rectangle; in fact, any polygon whose edges don't cross is acceptable.) The two approaches are equivalent, but the second is more convenient for present purposes.

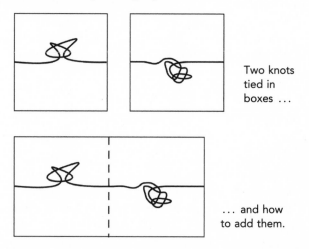

Two knots tied in boxes ...

... and how to add them.

If you tie two knots K and L in turn along two separate strings, then they can be 'added' by joining the ends together. Call the result $K + L$. The *unknot*, a straight string without a knot, can sensibly be denoted by 0, because $K + 0$ is topologically equivalent to K, which we can write as $K + 0 = K$ by employing the equals sign to indicate topological equivalence. The usual algebraic rules

$$K + L = L + K, \qquad K + (L + M) = (K + L) + M$$

can also be proved; the second one is easy, the first requires more thought.

Now we can see why Whodunni's trick must, indeed, be a trick. In effect, he *appeared* to tie two knots K and $K^\star$ that cancelled each other out. Now, if two knots K and $K^\star$ cancel, then

$$K + K^\star = 0 = K^\star + K$$

I'm tempted to replace K^* by $-K$, because it plays the same role, but the notation gets a bit messy if I do.

The outrageous idea is to consider the *infinite* knot

$$K + K^* + K + K^* + K + K^* + \cdots$$

Bracketed like this:

$$(K + K^*) + (K + K^*) + (K + K^*) + \cdots$$

we get $0 + 0 + 0 + \cdots$, which in topology as well as arithmetic is equal to 0. But bracketed like *this*:

$$K + (K^* + K) + (K^* + K) + (K^* + K) + \cdots$$

we get $K + 0 + 0 + 0 + \cdots$, which in topology as well as arithmetic is equal to K. Therefore, $0 = K$, so K was not a genuine knot to begin with.

In the previous item, we saw that this argument is not legitimate for numbers, and that's what makes the proof seem outrageous. However, with some technical effort it turns out that it *is* legitimate for knots. You just have to define the infinite 'sum' of knots using ever-smaller boxes. If you do that, the sum converges to a well-defined knot. The manipulations with brackets are correct. I don't claim that's obvious, but if you're a topologist it pretty much is.

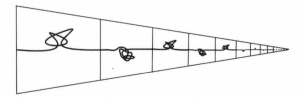

Tying a wild knot inside a triangle formed from an infinite sequence of shrinking trapezoidal boxes.

Infinite knots like this are called wild knots, and as the name suggests they should be handled with care. A mathematician called Raymond Wilder invented an especially unruly class of knots. You can guess what those are called.

Colorado Smith and the Solar Temple

Smith and Brunnhilde had penetrated to the inner sanctum of the Solar Temple of Psyttakosis IV, overcoming various minor obstacles on the way, such as the Pit of Everlasting Flame, the Creepy Crocodile Corridor, and the Valley of Vicious Venomous Vipers. Now, panting slightly from their exertions, they stood at the edge of the Temple Plaza – a square array of 64 slabs, four of which were decorated with a golden sun-disc. Behind them, the only entrance had been closed by a shining disc of solid gold with the weight of a dozen elephants.

But that was only to be expected. As Smith said, 'We'll just have to think our way out.'

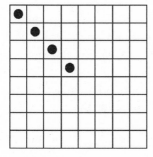

Location of
the sun-discs.

For once, Brunnhilde did not find this entirely reassuring. Maybe it was the earthquake and the puffs of dust thickening the air around them. Or was it the roar of approaching water? The carpet of scorpions on the floor, scuttling from cracks in the stonework? Or just the spikes round all the walls, which even now were extending towards them?

'What do we have to do *this* time?' she asked, having been in this position so often that she knew the script by heart.

'According to the Lost Papyrus of Bentnosy, we must choose four non-overlapping connected regions, each composed of 16 slabs, so that each includes one slab with a sun-disc,' replied Smith. 'Then the secret exit will open and let us into the adjoining treasure chamber – the one with those caskets of diamonds and emeralds I told you about. From there we just have to get through the underground maze that leads to the—'

'That seems easy enough,' said Brunnhilde, quickly sketching a solution. She caught his eye. 'What's the catch, Smith?'

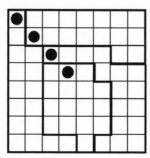

Not like this!

'Ah ... Well, according to an obscure inscription on the Oxyrhincus Ostracon of Djamm-Ta'art, which is a Late Period commentary on Bentnosy's papyrus, each of the four regions must be the same shape.'

'Ah. That makes it harder.' Brunnhilde smiled a hopeful smile, and tore up her sketch. 'I suppose the answer is in Bentnosy's papyrus?'

'Apparently not,' said Smith. 'It's not on the Ostracon, either – front or back.'

'Oh. Well, do you think we'll be able to work it out before that huge block of granite squashes us to the thickness of gold leaf?'

'What block of granite?'

'The one hanging over our heads on burning ropes.'

'Oh, *that* block of granite. Strange, Bentnosy didn't mention anything like that.'

Help Smith and Brunnhilde escape their dire predicament.

Answer on page 308

Why Can't I Add Fractions Like I Multiply Them?

Well, you can if you wish – it's a free country. Allegedly. But you won't get the right answer.

At school we are taught an easy way to multiply fractions: just multiply the numbers on the top, and those on the bottom, like this:

$$\frac{2}{5} \times \frac{3}{7} = \frac{2 \times 3}{5 \times 7} = \frac{6}{35}$$

But the rule for adding them is much messier: 'Put them over the same denominator (bottom), then add the numerators (tops).' Why can't we add them in a similar way? Why is

$$\frac{2}{5} + \frac{3}{7} = \frac{2+3}{5+7} = \frac{5}{12}$$

wrong? And what should we do instead?

Answer on page 308

Answer on page 308

• •

Farey, Farey, Quite Contrary

As soon as you say that some mathematical idea makes no sense, it turns out to be really useful and perfectly sensible. Although the rule

$$\frac{a}{b} + \frac{c}{d} = \frac{a+b}{c+d}$$

is not the correct way to add fractions, it is still a possible way to *combine* them, as the geologist John Farey, Sr, suggested in 1816 in the *Philosophical Magazine*. He hit on the idea of writing all fractions a/b whose denominator b is less than or equal to some specific number, in numerical order. Only fractions whose numerical values lie between 0 and 1 (inclusive) are allowed, so $0 \le a \le b$. To avoid repetitions, he also required the fraction to be in 'lowest terms', which means that a and b do not have a common factor (bigger than 1). That is, a fraction like $\frac{4}{6}$ is disallowed, because 4 and 6 both have the common factor 2. It should be replaced by $\frac{2}{3}$, which has the same numerical value but doesn't involve common factors.

The resulting sequences of fractions are called Farey sequences. Here are the first few:

$b \leq 1$: $\quad \frac{0}{1} \quad \frac{1}{1}$

$b \leq 2$: $\quad \frac{0}{1} \quad \frac{1}{2} \quad \frac{1}{1}$

$b \leq 3$: $\quad \frac{0}{1} \quad \frac{1}{3} \quad \frac{1}{2} \quad \frac{2}{3} \quad \frac{1}{1}$

$b \leq 4$: $\quad \frac{0}{1} \quad \frac{1}{4} \quad \frac{1}{3} \quad \frac{1}{2} \quad \frac{2}{3} \quad \frac{3}{4} \quad \frac{1}{1}$

$b \leq 5$: $\quad \frac{0}{1} \quad \frac{1}{5} \quad \frac{1}{4} \quad \frac{1}{3} \quad \frac{2}{5} \quad \frac{1}{2} \quad \frac{3}{5} \quad \frac{2}{3} \quad \frac{3}{4} \quad \frac{4}{5} \quad \frac{1}{1}$

Farey noticed – but could not prove – that in any such sequence, the fraction immediately between a/b and c/d is the 'forbidden sum' $(a + b)/(c + d)$. For instance, between $\frac{1}{2}$ and $\frac{2}{3}$ we find $\frac{3}{5}$, which is $(1 + 2)/(2 + 3)$. Augustin-Louis Cauchy supplied a proof in his *Exercises de Mathématique*, crediting Farey with the idea. Actually, it had all been published by C. Haros in 1802, but nobody had noticed.

So, although you can't *add* two fractions this way, the formula has its uses, and we can define the *mediant*

$$\frac{a}{b} \oplus \frac{c}{d} = \frac{a+b}{c+d}$$

provided that the fractions are in lowest terms. One problem with them not being in lowest terms is that different versions of the same fraction can lead to different results. For example,

$$\frac{1}{2} \oplus \frac{1}{3} = \frac{2}{5} \quad \text{but} \quad \frac{1}{2} \oplus \frac{2}{6} = \frac{3}{8}$$

which is different.

Farey sequences are widely used in number theory, and also show up in non-linear dynamics – 'chaos theory'.

● ●

Pooling Resources

Alice and Betty owned adjacent market stalls, and both were selling cheap plastic bracelets. Each had 30 bracelets. Alice had decided to price hers at 2 for £10, while Betty was thinking of charging 3 for £20. So together they would make £150 + £200 = £350, provided that they both sold all their bracelets.

Worried that the competition might destabilise the market, they decided to pool their resources, and reasoned that 2 for £10 and 3 for £20 combine to give 5 for £30. At that price, if they sold all 60 bracelets, then their total income would be £360, which was £10 better.

Just across the way, Christine and Daphne were also selling bracelets, and also had 30 each to sell. Christine was thinking of selling hers at 2 for £10, while Daphne was thinking of undercutting the competition severely by selling hers at 3 for £10. When they got wind of what Alice and Betty were doing, they too decided to pool their resources, and sell their 60 bracelets at 5 for £20.

Was this a good idea?

Answer on page 310

Welcome to the Rep-Tile House

A *rep-tile* is more properly known as a replicating polygon, and it is a shape in the plane that can be dissected into a number of identical copies, each the same shape but smaller. The shapes are allowed to have their boundaries in common, but do not otherwise overlap. If the polygon has s sides and it dissects into c copies, it is called a c-rep s-gon. Several different 4-sided rep-tiles (4-gons) are known. Most are 4-rep, but there are k-rep 4-gons for every k.

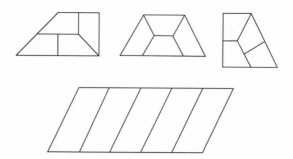

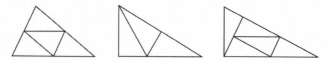

Top: replicating 4-gons. If the parallelogram at the bottom has sides 1 and √k, then it is rep-k.

Every triangle (3-gon) is 4-rep. Some special triangles are 3-rep or 5-rep.

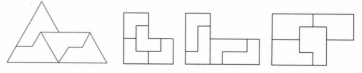

Replicating 3-gons. The first can be any shape. The second has sides 1 (vertical) and √3 (horizontal). The third has sides 1 (vertical) and 2 (horizontal).

Only one 5-sided rep-tile has yet been discovered: the *sphinx*. It requires four copies. There is a unique 5-rep 3-gon (triangle), and exactly three 4-rep 6-gons are known.

The only 4-rep 5-gon, the sphinx, and the three known 4-rep 6-gons.

There are several rep-tiles that stretch 'polygon' to the limit. And some go beyond that, having infinitely many sides – but, hey, let's be broad-minded.

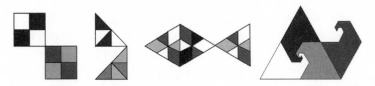

More exotic rep-tiles.

The first 4-rep 4-gon in the first picture is also rep-9. Can you dissect it into nine copies of itself? As far as I am aware, every known rep-4 tile is also rep-9, but this has not been proved in general.

Answer on page 310

Cooking on a Torus

Now, I'm going to set the utilities puzzle (*Cabinet*, page 199; and *Hoard*, page 117) for the third time, with a new twist. Metaphorically and literally. Three houses have to be connected to three utility companies – water, gas, electricity. Each house must be connected to all three utilities. Can you do this without the connections crossing? Assume there is no third direction to pass pipes over or under cables, and you are not allowed to pass the connections through a house or a utility company building. Note: *connections*. No quibble-cooks (see page 116) allowed!

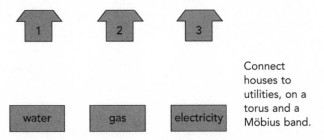

Connect houses to utilities, on a torus and a Möbius band.

What's the difference this time? I'm not asking you to work in the *plane*. Try it on a torus (metaphorical twist) and a Möbius band (literal twist). A torus is a surface with a hole, like a doughnut. A Möbius band is formed by joining the ends of a strip of paper with a half-twist (*Cabinet,* page 111).

By the way: mathematicians think of a surface like the Möbius band as having zero thickness, so that the utilities, houses and lines connecting them lie *in* it, not *on* it. But a real sheet of paper actually has two distinct surfaces, very close together. You can either think of the surface as being transparent, or (better) imagine that the lines are drawn on paper with ink that soaks through, so that everything is visible on both surfaces of the paper.*

If you don't use this convention, then some of the lines in my answer end up on the back of the strip and don't join up with the houses or utilities. You are then trying to solve the analogous problem on a cylindrical band with a double-twist. Topologically, this is the same as an ordinary cylindrical band, and in particular it has two distinct sides. Now there is no solution. Why not? A cylinder can be flattened out topologically in the plane, to form an *annulus* – the region between two circles. So any solution of the puzzle on a cylindrical band also provides a solution in the plane. But no solution in the plane exists without cooking (*Cabinet,* page 199).

Answers on page 311

• •

The Catalan Conjecture

Anyone who plays around with numbers soon notices that the consecutive integers 8 and 9 are both perfect powers (higher than the first power, of course). In fact, 8 is 2-cubed and 9 is 3-squared.

---

* Except when checking that it has only one side, by colouring it. Then the ink *doesn't* soak through. If it did, an ordinary cylinder would have one side. Because of difficulties like this, mathematicians approach the whole topic differently, talking of 'orientations' rather than 'sides'.

Are there any other positive whole numbers with this property – consecutive or not? (Powers higher than the cube are permitted, and strictly speaking 0 is not positive: it is non-negative. So this rules out $1^m - 0^n = 1$.) In 1844, the Belgian mathematician Eugène Catalan conjectured that the answer is no – that is, the Catalan equation

$$x^a - y^b = 1$$

has only the above solutions in positive integers x and y when a and b are integers ≥ 2. In a mathematical publication known as Crelle's Journal,[*] he wrote: 'Deux nombres entiers consécutifs, autres que 8 et 9, ne peuvent être des puissances exactes; autrement dit: l'équation $x^m - y^n = 1$, dans laquelle les inconnues sont entières et positives, n'admet qu'une seule solution.'

The problem has a long history. The composer Philippe de Vitry (1291–1361) stated that the only powers of 2 and 3 that differ by 1 are (1,2), (2,3), (3,4) and (8,9). Levi ben Gerson (1288–1344) provided a proof that de Vitry was right: $3^m \pm 1$ always has an odd prime factor if $m > 2$, so it cannot be a power of 2. By 1738, Euler had completely solved the equation $x^2 - y^3 = 1$ in whole numbers, proving that the only positive solution is $x = 3$, $y = 2$. But Catalan's conjecture allows higher powers than the cube, so these earlier results were not sufficient to prove it.

In 1976, Robert Tidjeman proved that Catalan's equation has only finitely many solutions; indeed, any solution must have $x, y < \exp \exp \exp \exp 730$, where $\exp x = e^x$. However, this upper limit on the size is almost inconceivably gigantic – and in particular far too large for a computer search to eliminate all other potential solutions. In 1999, M. Mignotte proved that in any hypothetical solution, $a < 7.15 \times 10^{11}$ and $b < 7.78 \times 10^{16}$, but the gap is still too big for a computer to fill. A solution

---

[*] More properly the *Journal für die reine und angewandte Mathematik* (Journal for Pure and Applied Mathematics).

seemed hopeless. But, in 2002, the mathematical world was stunned when the Romanian-born German mathematician Preda Mihailescu proved that Catalan was right, with a clever proof based on cyclotomic numbers – complex nth roots of 1. So the conjecture has now been renamed Mihailescu's theorem.

There is a generalisation of the problem, to so-called Gaussian integers, which are complex numbers $p + q$i, where p and q are ordinary integers and i $= \sqrt{-1}$. Here, there exist two non-trivial powers that differ not by 1, but by i:

$$(78 + 78\text{i})^2 - (-23\text{i})^3 = \text{i}$$

As far as I know, the corresponding conjecture – that this or minor variations are the only new cases where two Gaussian integer powers differ by 1, −1, i, or −i – remains open.

An extensive history of the problem can be found at: www.math.leidenuniv.nl/~jdaems/scriptie/Catalan.pdf

• •

The Origin of the Square Root Symbol

The square root symbol

has a wonderfully arcane look, like something out of an ancient manuscript on alchemy. It's the sort of symbol wizards would write, and formulas that contain it always look impressive and mysterious. But where did it come from?

Before 1400, European writers on mathematics generally used the word 'radix' for 'root' when referring to square roots. By the late medieval period, they abbreviated the word to its initial letter, a capital R with a short stroke through it:

The Renaissance Italian algebraists Girolamo Cardano, Luca

Pacioli, Rafael Bombelli and Tartaglia (Niccolò Fontana) all used this symbol.

The symbol $\sqrt{}$ is in fact a distorted letter *r*. How mundane! It first appeared in print in the first German algebra text, Christoff Rudolff's *Coss* of 1525, but it took several centuries to become standard.

The site
www.roma.unisa.edu.au/07305/symbols.htm
discusses the history of many other mathematical symbols.

• •

Please Bear with Me

Q: What's a polar bear?

A: A Cartesian bear after a change of coordinates.

• •

The Ham Sandwich Theorem

I'm not making this up: that's what it is called. It says that if you make a ham sandwich from two slices of bread and a slice of ham, then it is possible to cut the sandwich along some plane so that each of these three components is divided exactly in half, by volume.

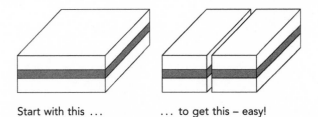

Start with this ...   ... to get this – easy!

This is fairly obvious if the bread and the ham form nice square slabs, neatly arranged. It is less obvious if you appreciate that mathematicians are referring to *generalised* bread and ham, which may take any shape whatsoever. (One immediate consequence is the cheese sandwich theorem, which might otherwise need a separate proof. Generality and power go hand in hand.)

A mathematician's ham sandwich.

There are some technical conditions: in particular, the three pieces must not be so terribly complicated that they don't *have* well-defined volumes (see *Cabinet*, page 163). In compensation, there is no requirement for a 'piece' to be connected – all in one lump, so to speak – but if it's not you only have to divide the overall lump in half, not each separate part of it. Otherwise you're trying to prove the ham and cheese sandwich theorem, which is false – see below.

The ham sandwich theorem is actually quite tricky to prove, and it is mostly an exercise in topology. To give you a flavour of the proof, I'll show you how to deal with the simpler case of two shapes in two dimensions – the Flatland cheese-on-toast theorem.

Here's the problem:

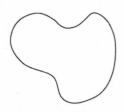 

Find a line that splits both cheese (white) and toast (grey) in half, by area.

Here's how to prove it can be solved. Pick a direction and find a line pointing that way, which splits the cheese in half. It is not hard to prove that precisely one such line exists.

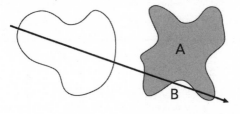

Start with a line in some direction (shown by the arrow) that splits the cheese in half.

Of course, unless you've got lucky, this line won't split the toast in half too, but there will be two parts A and B on opposite sides of the line, with A on the left and B on the right if you look along the arrow. (Here B includes *both* chunks of the toast on that side. In general, either A or B might be empty – that doesn't change the proof.) Suppose that, as shown, A has a bigger area than B.

Now gradually rotate the direction you're thinking of, and do the same thing for each new direction.

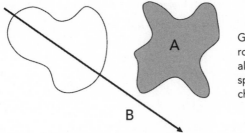

Gradually rotate the line, always splitting the cheese in half.

Eventually you will have rotated the direction by 180°. Since only one line splits the cheese in half, this line must coincide with the original one, except that the arrow now points the other way:

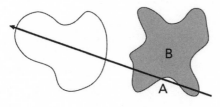

After a 180° rotation, the line has exactly reversed direction, and regions A and B have swapped places.

Because the arrow points the other way, parts A and B of the toast have changed places. At the start, A was bigger than B, so now B must be bigger than A. (The pieces are the same as they were at the start, only the labels A and B have swapped.) However, the areas of A and B vary *continuously* as the angle of the line rotates. (This is where the topology comes in.) Since initially area(A) > area(B) and finally area(A) < area(B), there must be some angle in between for which area(A) = area(B). (Why? The difference area(A) − area(B) also varies continuously, starts positive, and ends negative. Somewhere in between it must be zero.) This proves the Flatland cheese-on-toast theorem.

This type of proof doesn't work in three dimensions, but the theorem is still true. It seems to have first been proved by Stefan Banach, Hugo Steinhaus, and others in 1938. A version about simultaneously bisecting *n* pieces in *n* dimensions was proved by Arthur Stone and John Tukey in 1942.

Here are two easier puzzles for you, which explain some of the limitations:

- Show that it is not always possible to bisect *three* regions of the plane with a single straight line.
- Show that the ham and cheese sandwich theorem is false: it is not always possible to bisect *four* regions of space with a single plane.

Answers on page 311

More on this theorem, and an outline of the proof, can be found at:
en.wikipedia.org/wiki/Ham_sandwich_theorem

Cricket on Grumpius

On planet Earth, and in those countries that play the game,* cricket fans always get very upset when a batsman scores 49 and then gets out, because he has just missed a half-century. But this is a horribly decimalist way of viewing the situation.

The inhabitants of the distant alien world of Grumpius are a case in point. Oddly enough, when humans first made contact with their civilisation, it turned out that they were passionate about cricket. Astrobiologists speculate that the Grumpians must have picked up our satellite TV programmes during an exploratory trip through the Solar System.

Out for 49 –
congratulations!

Anyway, whenever a Grumpian batsthing scores what we would write as 49, the crowd goes wild, and the batsthing raises its bat and bobs its tentacles in the Grumpian equivalent of a punched fist. Why?

Answer on page 312

• •

---

* Whose total population far exceeds that of the world's baseball-playing countries.

The Man Who Loved Only Numbers

The brilliant Hungarian mathematician Paul Erdős was distinctly eccentric. He never held a formal academic position, and he never owned a house. Instead, he travelled the world, living for short periods with his colleagues and friends. He published more collaborative papers than anyone else, before or since.

He knew the phone numbers of many mathematicians by heart, and would phone them anywhere in the world, ignoring local time. But he could never remember anyone's first name – except for Tom Trotter, whom he always addressed as Bill.

One day, Erdős met a mathematician. 'Where are you from?', he enquired.

'Vancouver.'

'Really? Then you must know my friend Elliot Mendelson.'

There was a pause. 'I *am* your friend Elliot Mendelson.'

Paul Erdős.

The Missing Piece

'Ooooh! Jigsaws!' yelled Innumeratus. 'I *love* jigsaws!'

'This one is special,' said Mathophila. 'There are 17 pieces, forming a square. I've laid them out on a square grid, and every corner of every piece lies exactly on the grid.'

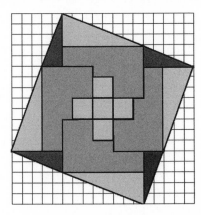

Rearrange the pieces to form the *same* square... with one piece left over.

'Now,' she continued, 'I'm going to take away one of the small squares, and your job is to fit the other 16 pieces back together again to make the same big square that we started with.'

Innumeratus saw no contradiction in that, and half an hour later he proudly showed his answer to Mathophila.

What was his answer, and how can he form the same square when one piece is missing? [*Hint*: it can't really be exactly the same. And maybe that initial 'square' isn't actually square ...]

Answer on page 312

● ●

The Other Coconut

A mathematician and an engineer are marooned on a desert island, which has two palm trees: one very tall, the other much shorter. Each has one coconut, at the very top.

The engineer decides to have a try for the more difficult

coconut, on top of the tall tree, while they still have the energy to reach it. He clambers up, scraping his legs raw, and eventually returns with the coconut. They smash it open with a rock and eat and drink the contents.

Three days later, both of them now weak with hunger and thirst, the mathematician volunteers to get the other coconut. He climbs the shorter tree, detaches the coconut, and brings it down. The engineer then watches bemusedly while the mathematician starts climbing the taller tree, groaning and sweating profusely, finally gets to the top, deposits the coconut there, and makes his way back down with even more difficulty. He is completely exhausted.

The engineer stares at him, then up at the distant coconut, and then back to the mathematician. 'Whatever possessed you to do *that*?'

The mathematician glares back. 'Isn't it obvious? I've reduced it to a problem that we already know how to solve!'

• •

What Does Zeno?

Zeno of Elea was an ancient Greek philosopher who lived around 450 BC, and he is best known for *Zeno's Paradoxes* – four thought experiments, each of which aims to prove that motion is impossible. Some of them may not have originated with Zeno, and others may not even have been stated by Zeno – the evidence is debatable – but I'll list the traditional four, starting with the best known:

Achilles and the Tortoise

These two characters agree to have a race, but Achilles can run faster than the tortoise, so he gives the creature a head start. The tortoise argues that Achilles can never catch him, because by the time Achilles has reached the position where the tortoise *was*, it has moved ahead. And by the time Achilles has reached *that* position, the tortoise has moved ahead again … So Achilles has to pass through infinitely many locations before he can catch up, which is impossible.

Achilles in hot pursuit.

The Dichotomy

In order to reach some distant location, you must first reach the halfway mark, and before you do that, you must reach the quarterway mark, and before that ... So you can't even get started.

The Arrow

At any instant of time, a moving arrow is stationary. But if it is always stationary, it can't move.*

The Stadium

This one is more obscure. Aristotle refers to it in his *Physics*, and says roughly this: 'Two rows of bodies, each composed of an equal number of bodies of equal size, pass each other on a racecourse, proceeding with equal velocity in opposite directions. One row originally occupies the space between the goal and the middle point of the course; the other that between the middle point and the starting post. The conclusion is that half a given time is equal to double that time.'

---

* In Terry Pratchett's Discworld novel *Pyramids*, there is an Ephebian philosopher named Xeno, who proved that an arrow cannot hit a running man. Other philosophers agreed, with the proviso that 'it is fired by someone who has been in the pub since lunchtime'. Xeno also claimed that the tortoise is the fastest animal on the Disc, but actually it is the ambiguous puzuma, which travels close to the speed of light. If you see a puzuma, it's not there.

What Zeno had in mind here is not at all clear.

Set-up for the stadium paradox.

As a practical matter, we know that motion *is* possible. While the tortoise is expounding his argument, Achilles shoots past him, oblivious to the impossibility of doing infinitely many things in a finite time. The deeper issue is: what *is* motion, and how does it happen? This question is about the physical world, whereas Zeno's paradoxes are about mathematical models of the real world. If his logic were correct, it would dispose of several possible models. Is it correct, though?

Most mathematicians and school mathematics teachers resolve (that is, explain away) the first two paradoxes by doing a few calculations. For instance, suppose that the tortoise moves at 1 metre per second, while Achilles moves at 10 metres per second. Start with the tortoise 100 metres in front. Tabulate the events that Zeno considered:

Time	Achilles	Tortoise
0	0	100
10	100	110
11	110	111
11.1	111	111.1
11.11	111.1	111.11

The list is infinitely long – but why worry about that? Where is Achilles after, say, 12 seconds? He has reached the 120-metre mark. The tortoise is behind, at 112 metres. Indeed, Achilles gets level with the tortoise after exactly $11\frac{1}{9}$ seconds, because at that instant, both of them have reached the $111\frac{1}{9}$-metre position.

As a follow-up, we might add that the infinite sequence

$$10, \quad 11, \quad 11.1, \quad 11.11, \quad 11.111, \quad \ldots$$

converges to $11\frac{1}{9}$, meaning that it approaches indefinitely close to that value, and that value alone, if you go far enough along the sequence.

The dichotomy paradox can be approached in a similar way. Suppose that the arrow has to travel 1 metre and moves at 1 metre per second. Zeno tells us where the arrow is after $\frac{1}{2}$ second, $\frac{1}{4}$ second, $\frac{1}{8}$ second, and so on. At none of these times has it reached its target. But that doesn't imply that there is *no* time at which it reaches the target – just that it's not one of those considered by Zeno. It doesn't get there after $\frac{2}{3}$ seconds, either, for instance. And it clearly *does* get there after 1 second.

And here, too, we can point out that the infinite sequence

$$1, \quad \frac{1}{2}, \quad \frac{1}{4}, \quad \frac{1}{8}, \quad \cdots$$

converges to zero, and the corresponding sequence of times

$$0, \quad \frac{1}{2}, \quad \frac{3}{4}, \quad \frac{7}{8}, \quad \cdots$$

converges to 1, the instant at which the arrow hits the target.

Many philosophers are less satisfied with these resolutions than mathematicians, physicists and engineers are. They argue that these 'limit' calculations do not explain why infinitely many different things can happen in turn. Mathematicians tend to reply that they show *how* infinitely many different things can happen in turn, so the assumption that they can't is what's making everything seem paradoxical. When the arrow travels from the 0-metre mark to the 1-metre mark, it does so in the finite time of 1 second. But although the length of the interval from 0 to 1 is finite, the number of *points* in it (in the usual 'real number' model) is infinite. In such a model, *all* motion involves passing through infinitely many points[*] in a finite time.

I don't claim that my discussion knocks the argument on the

---

[*] Indeed, a continuum, which according to Cantor is a bigger kind of infinity than that of the whole numbers (see *Cabinet*, page 160).

head, or covers all relevant points of view. It's just a quick and broad summary of a few of the main issues.

The arrow paradox is also often resolved by taking the 'limit' point of view, or, more precisely, calculus, which is what limits were invented for. In calculus, a moving object can have an instantaneous speed that is not zero, even though it has a fixed location at that instant. Making logical sense of this took a few centuries, and boils down to taking the limit of the average speed over shorter and shorter intervals of time. Again, some philosophers feel that this is not an acceptable approach.

I think there's another interesting mathematical point buried in this one. Physically, there is a definite difference between an arrow that is moving and one that is not, even if they are both in the same place at some instant. The difference can't be seen in an instantaneous 'snapshot', but nevertheless it is physically real (whatever that means). Anyone who does classical mechanics knows what the difference is: a moving body has *momentum* (mass times velocity). A snapshot tells you the position of the body, but not its momentum. These are independent variables: in principle, a body can have any position and any momentum.

While position is directly observable (see where the body is), momentum is not. The only way we know to observe it is to measure the velocity, which involves at least two positions, at closely spaced intervals of time. Momentum is a 'hidden variable', whose value must be inferred indirectly. Since 1833, the most popular formulation of mechanics has been the one proposed by Sir William Rowan Hamilton, which explicitly works with these two kinds of variables, position and momentum. So the difference between a moving arrow and a fixed one is that the moving one has momentum, whereas the fixed one does not. How can you tell the difference? Not by taking a snapshot. You have to wait and see what happens next. The main thing missing from this approach, philosophically, is any description of what momentum *is*, physically. And that's probably a lot harder than anything that worried Zeno.

What of the stadium? One answer is that Zeno was hopelessly

confused, and that his conclusion 'half the time is equal to double the time' does not follow from his set-up. But there is an interpretation that puts all four paradoxes in a more interesting light. The suggestion is that Zeno was trying to understand the nature of space and time.

The most obvious models of space are either that it is discrete, with isolated points placed at (say) integer positions 0, 1, 2, 3, and so on; or it is continuous, and points correspond to real numbers, which can be subdivided as finely as we wish. The same goes for time.

discrete

continuous

Possible structures for space and time.

Altogether, these choices give four distinct combinations for the structure of space and time. And these relate fairly convincingly to the four paradoxes, like this:

Paradox	Space	Time
Achilles and the tortoise	Continuous	Continuous
Dichotomy	Discrete	Continuous
Arrow	Continuous	Discrete
Stadium	Discrete	Discrete

Possibly Zeno was trying to show that each combination suffers from logical problems.

- The first requires infinitely many things to happen over a finite period of time.
- The second means that space cannot be subdivided indefinitely, while time can. So consider an object traversing the shortest possible unit of space, in some non-zero time t. At time 0, it is in one location; at time t, it is in the closest different location. So, where is it at time $\frac{1}{2}t$? It should be halfway between, but in this discrete version of space, there is no point in between.
- If space is continuous and time discrete, then the same thing

happens with time and space interchanged. The arrow manages to move from a fixed location at one instant to a different fixed location at the next. It could go in between, but there isn't a *time* in between for it to get there.

- What of the stadium? Now both space and time are discrete. So imagine Zeno's two rows of identical bodies passing each other. To clarify the problem, let's add a third row of bodies, which doesn't move, and compare each moving row with that. Assume that relative to the fixed row, they move as rapidly as possible: that is, each moves through the smallest possible unit of space in the smallest possible unit of time.

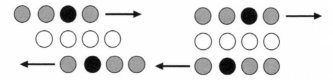

Successive positions of the rows of identical bodies.

You'll notice that I've made two of the bodies black: this is for reference. At the first instant, the black bodies are one unit of space apart, with the top one on the left. At the next instant, they are one unit of space apart, with the top one on the right: they have swapped positions.

At which instant were they level with each other?

They weren't. Because we are working with the smallest possible interval of time, what the pictures show is everything that happens. There is no 'halfway time' for the two black bodies to get level with each other. This problem is not insurmountable – we can just accept that a moving body makes this kind of 'jump', for instance. And maybe the whole neat and tidy classification of the paradoxes into four possibilities is misleading, and Zeno's intentions were quite different.

Pieces of Five

'Here's a fine challenge for ye, me hearties!' yelled Roger Redbeard, the pirate captain, who liked to keep his crew's minds alert. If only to check they still had them.

He held up four coins, identical gold pieces-of-eight.

'Now, me lads – what I wants ye to do is to place these four gold coins so that they be *equidistant*.'

Seeing the baffled looks on their faces, he explained. 'What I means, lads, is that the shortest distance between any two coins 'as to be the same as that between any other two coins.'

To his considerable surprise, the bosun immediately realised that it was no good 'working in the plane', and the solution required three dimensions of space. He quickly found an answer: place three of the coins touching each other in a triangle, and sit the fourth on top. All coins are touching, so all distances between them are zero, hence equal.

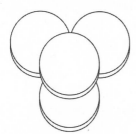

How to do it
with four coins.

Redbeard, dismayed, thought for a moment. 'So, ye think ye be smart? Try doin' it with *five* coins, then. Make *them* all equidistant from each other!'

Eventually the bosun found an answer, but it wasn't easy. What was it?

Answer on page 313

Pi in the Sky

It is not widely known that you can work out the value of π by observing the stars. Moreover, the reasoning behind this feat is not based on astronomy, but on number theory – and it works, not because of a pattern in the stars, but because there isn't one.

Suppose you pick two non-zero whole numbers at random, less than or equal to some upper limit. The probability should be uniform – that is, each number should have the same chance of being chosen. For instance, the upper limit might be a million, and the numbers you get might be 14,775 and 303,254, say, each with probability one in a million. Now ask: Do those two numbers have a common factor (greater than 1) – or not? In this case they don't. In general, number-theorists have proved that the proportion of pairs with no common factor tends to $6/\pi^2$ as the upper limit becomes arbitrarily large. This remarkable result is one of many properties of π that appear to have no connection to circles. It is exact, not an approximation, and it can (with some clever tricks) be deduced from the formula

$$1 + \frac{1}{4} + \frac{1}{9} + \frac{1}{16} + \frac{1}{25} + \cdots = \frac{\pi^2}{6}$$

In 1995, Robert Matthews wrote a short letter to the scientific journal *Nature*, pointing out that this theorem in number theory can be used to extract a reasonably accurate value of π from the stars in the night sky – on the assumption that the positions of the stars are random. His idea was to work out the angular distances between lots of stars (that is, the angle between the lines joining those stars to the observer's eye) and then to transform those distances into large integers. (The actual formula he used was to take the cosine of the angle, add 1, and multiply by half a million.) If you ignore anything after the decimal point, and exclude zero, you get a list of positive integers between 1 and a million. Pick pairs at random, and let the proportion with no highest common factor be p. Then p is approximately $6/\pi^2$, so π is approximately $\sqrt{6/p}$.

Matthews did this for the 100 brightest stars in the sky, producing a list of 4,095 integers between 1 and a million. From these he derived a million pairs of randomly chosen numbers, and found that $p = 0.613333$. Thus π should be approximately 3.12772. This isn't as good as the school approximation 22/7, but it is within 0.4 per cent of the correct value. Using more stars should improve it. Matthews ended his letter by saying that 'Latter-day Pythagoreans may take encouragement from learning that a 99.6 per cent accurate value for π can be found among the stars over their heads.'

The Curious Incident of the Dog

In Sir Arthur Conan Doyle's Sherlock Holmes story 'Silver Blaze', we find:

'Is there any other point to which you would wish to draw my attention?'

'To the curious incident of the dog in the night-time.'

'The dog did nothing in the night-time.'

'That was the curious incident,' remarked Sherlock Holmes. Here is a sequence:

1, 2, 4, 7, 8, 11, 14, 16, 17, 19, 22, 26, 28, 29, 41, 44

Having taken Holmes's point on board: what is the next number in the sequence?

Answer on page 313

Answer on page 313

Mathematics Made Difficult

There's a snag with all these 'find the next number in the sequence' puzzles – the answer need not be unique. Carl Linderholm put his finger on this problem in the often-hilarious spoof *Mathematics Made Difficult*, published in 1971 when the 'new math(s)' was in vogue. In it, he remarks: 'Mathematicians always strive to confuse their audiences; where there is no confusion, there is no prestige.' As an example, Linderholm defines the natural number system as a 'universal pointed function'.

His take on 'guess the next number' puzzles is unusual but logical. For example, to find the next number after

8, 75, 3, 9

he tells you to write down 'the only answer any sensible person would put there'. Which is – what? Ah, that's the clever part. As a clue, here are some more puzzles of the same type:

- What comes after 1, 2, 3, 4, 5?
- What comes after 2, 4, 6, 8, 10?
- What comes after 1, 4, 9, 16, 25?
- What comes after 1, 2, 4, 8, 16?
- What comes after 2, 3, 5, 7, 11?
- What comes after 139, 21, 3, 444, 65?

Here are the answers we would obtain using Linderholm's method:

- 19
- 19
- 19
- 19
- 19
- 19

What is the justification for this bizarre set of answers? It is Lagrange's interpolation formula, which provides a polynomial $p(x)$ such that $p(1)$, $p(2)$, ..., $p(n)$ is any specified sequence of length n, for any finite n. *Some* such p must fit the sequence

1, 2, 3, 4, 5, 19

so the choice of 19 is justified by the polynomial. The same goes for all the other examples. As Linderholm explains, this answer is far superior to

1, 2, 3, 4, 5, 6

because his procedure 'is much the simpler, and is easier to use, and is obtained by a more general method'.

Why 19? Choose your favourite number and add 1. Why add 1? To 'make it more difficult to determine your character defects by analysing your favourite number. No technique by which a person's character may be found out from his secret number is known to the author, but of course someone may some day invent such a technique.'

In the spirit of Linderholm's book, I really ought to show you Lagrange's interpolation formula. So it's on page 313.

. . . .

A Weird Fact about Egyptian Fractions

Ron Graham has proved that any number greater than 77 can be expressed as a sum of distinct positive integers, whose reciprocals (1 divided by the appropriate integer) add up to 1. So this represents 1 as an Egyptian fraction (see page 76).

For example, let $n = 425$. Then

$$\frac{1}{3} + \frac{1}{5} + \frac{1}{7} + \frac{1}{9} + \frac{1}{15} + \frac{1}{21} + \frac{1}{27} + \frac{1}{35} + \frac{1}{63} + \frac{1}{105} + \frac{1}{135} = 1$$

and $3 + 5 + 7 + 9 + 15 + 21 + 27 + 35 + 63 + 105 + 135 = 425$.

On the other hand, Derrick Henry Lehmer showed that 77 cannot be written in this form. So here we have a special property of the number 77, in the context of Egyptian fractions.

. . . .

A Four Colour Theorem

If I arrange three equal circles so that each touches the other two, then it's obvious that I need three colours if I wish to colour each circle so that circles that touch have different colours. The picture shows three circles, each touching the other two, so they all need different colours.

Three colours
required.

Four equal circles in the plane can't all touch each other, but that doesn't mean three colours always work: there are more complicated ways to arrange lots of coins, and some of those might need four colours. What is the *smallest* number of equal circles that can be arranged so that four colours are needed? Again, the rule is that if two circles touch, they must have different colours.

Answer on page 314

Serpent of Perpetual Darkness

In 2004, astronomers discovered asteroid 99942, and named it Apophis after the ancient Egyptian serpent who attacks the Sun-god Ra during his nightly passage through the eternal darkness of the Underworld.* It was an appropriate name in some ways, because the astronomers also announced that there was a serious danger that the newly discovered asteroid might collide with the Earth on 13 April 2029 – or, if not, on 13 April 2036. The chance of a collision was initially estimated at 1 in 200, and peaked at 1 in 37, but is now thought to be highly unlikely.

A popular British journalist wrote, in his regular column, something to the effect 'How come they can be so specific about the date, but they don't know the year?' Now, to be fair, it was a humorous column, and it's quite an amusing question. But it has a serious answer.

---

* It is also the name of the principal villain in *Stargate SG-1*, a Goa'uld System Lord, in case that's more familiar.

Enlighten the journalist. [*Hint*: What is a year, astronomically speaking?]

Answer and discussion on page 315

What Are the Odds?

Mathophila takes a pack of cards, and places the four aces on the table, face down. So two (spades, clubs) are black; the other two (hearts, diamonds) are red.

Shuffle, place face down, pick two.

'Innumeratus?'

'Yes?'

'If you pick two of these cards at random, what is the probability that they have different colours?'

'Ummm ...'

'Well, either the colours are the same, or they're not, right?'

'Yes.'

'And there's the same number of cards of each colour.'

'Yes.'

'So the chances of your two cards being the same, or different, must be equal – so both are equal to $\frac{1}{2}$. Right?'

'Ummm ...'

Is Mathophila right?

Answer on page 316

A Potted History of Mathematics

*c.*23,000 BC	Ishango bone records the prime numbers between 10 and 20. Apparently.
*c.*1900 BC	Babylonian clay tablet Plimpton 322 lists what may be Pythagorean triples. Other tablets record movements of the planets and how to solve quadratic equations.
*c.*420 BC	Discovery of incommensurables (irrational numbers in geometric guise) by Hippasus of Metapontum.*
*c.*400 BC	Babylonians invent symbol for zero.
*c.*360 BC	Eudoxus develops a rigorous theory of incommensurables.
*c.*300 BC	Euclid's *Elements* makes proof central to mathematics, and classifies the five regular solids.
*c.*250 BC	Archimedes calculates the volume of a sphere, and other neat stuff.
*c.*36 BC	Mayans reinvent symbol for zero.
*c.*250	Diophantus writes his *Arithmetica* – how to solve equations in whole and rational numbers. Uses symbols for unknown quantities.
*c.*400	Symbol for zero re-reinvented in India. Third time lucky.
594	Earliest evidence of positional notation in arithmetic.
*c.*830	Muhammad ibn Musa al-Khwarizmi's *al-Jabr w'al-Muqabala* manipulates algebraic concepts as abstract entities, not just placeholders for numbers, and gives us the word 'algebra'. Doesn't use symbols, however.

---

* Hippasus was a member of the Pythagorean cult, and it is said that he announced this theorem while he and some fellow cultists were crossing the Mediterranean in a boat. Since Pythagoreans believed that everything in the universe is reducible to whole numbers, the others were less than overjoyed, and he was expelled. From the boat, according to some versions.

876	First undisputed use of a symbol for zero in positional base-10 notation.
1202	Leonardo's *Liber Abbaci* introduces the Fibonacci numbers through a problem about the progeny of rabbits. Also promotes Arabic numerals and discusses applications of mathematics to currency trading.
1500–1550	Renaissance Italian mathematicians solve cubic and quartic equations.
1585	Simon Stevin introduces the decimal point.
1589	Galileo Galilei discovers mathematical patterns in falling bodies.
1605	Johannes Kepler shows that the orbit of Mars is an ellipse.
1614	John Napier invents logarithms.
1637	René Descartes invents coordinate geometry.
*c.*1680	Gottfried Wilhelm Leibniz and Isaac Newton invent calculus and argue about who did it first.
1684	Newton sends Edmund Halley a derivation of elliptical orbits from the inverse square law of gravity.
1718	Abraham De Moivre writes first textbook on probability theory.
1726–1783	Leonhard Euler standardises notation such as e, i, π, systematises most known mathematics, and invents a huge amount of new mathematics.
1788	Joseph-Louis Lagrange's *Méchanique Analytique* places mechanics on an analytic basis, getting rid of pictures.
1796	Carl Friedrich Gauss discovers how to construct a regular 17-gon.
1799–1825	Pierre Simon de Laplace's five-volume epic *Mécanique Céleste* formulates the basic mathematics of the solar system.
1801	Gauss's *Disquisitiones Arithmeticae* provides a basis for number theory.

1821–1828 Augustin-Louis Cauchy introduces complex analysis.

1824–1832 Niels Henrik Abel and Évariste Galois prove that the quintic equation is not soluble using radicals; Galois paves the way for modern abstract algebra.

1829 Nikolai Ivanovich Lobachevsky introduces non-Euclidean geometry, followed shortly by János Bolyai.

1837 William Rowan Hamilton defines complex numbers formally.

1843 Hamilton formulates mechanics and optics in terms of the Hamiltonian.

1844 Hermann Grassmann develops multidimensional geometry.

1848 Arthur Cayley and James Joseph Sylvester invent matrix notation. Cayley predicts that it will never have any practical uses.

1851 Posthumous publication of Bernard Bolzano's *Paradoxien des Unendlichen* which tackles the mathematics of infinity.

1854 Georg Bernhard Riemann introduces manifolds – curved spaces of many dimensions – paving the way for Einstein's general relativity.

1858 Augustus Möbius invents his band.

1859 Karl Weierstrass makes analysis rigorous with epsilon–delta definitions.

1872 Richard Dedekind proves that $\sqrt{2} \times \sqrt{3} = \sqrt{6}$ – the first time this has been done rigorously – by developing the logical foundations of real numbers.

1872 Felix Klein's Erlangen programme represents geometries as the invariants of transformation groups.

c.1873 Sophus Lie starts working on Lie groups, and the mathematics of symmetry makes a huge leap forward.

1874	Georg Cantor introduces set theory and transfinite numbers.
1885–1930	Italian school of algebraic geometry flourishes.
1886	Henri Poincaré stumbles across hints of chaos theory and revives the use of pictures.
1888	Wilhelm Killing classifies the simple Lie algebras.
1889	Giuseppe Peano states his axioms for the natural numbers.
1895	Poincaré establishes basic ideas of algebraic topology.
1900	David Hilbert presents his 23 problems at the International Congress of Mathematicians.
1902	Henri Lebesgue invents measure theory and the Lebesgue integral in his PhD thesis.
1904	Helge von Koch invents the snowflake curve, which is continuous but not differentiable, simplifying an earlier example found by Karl Weierstrass and anticipating fractal geometry.
1910	Bertrand Russell and Alfred North Whitehead prove that $1+1=2$ on page 379 of volume 1 of *Principia Mathematica*, and formalise the whole of mathematics using symbolic logic.
1931	Kurt Gödel's theorems demonstrate the limitations of formal mathematics.
1933	Andrei Kolmogorov states axioms for probability.
c.1950	Modern abstract mathematics starts to take off. After that it gets complicated.

● ●

The Shortest Mathematical Joke Ever

Let $\varepsilon < 0$.

If you don't understand this one, see the note in the Answers section, page 317. If you do understand it and don't find it funny, congratulations.

● ●

Global Warming Swindle

Mathematical models are central to the study of global warming, because they help us understand how the Earth's atmosphere would behave with different levels of incoming radiation from the Sun, different levels of greenhouse gases such as carbon dioxide (CO_2) and methane, and whatever else might go into the model. I'll ignore the effect of methane – basically, it just makes everything worse. Climate change is a very complex topic, and this is just a quick look at one common misunderstanding.

Nearly all scientists working on climate are now convinced that human activities have increased the amount of CO_2 in the atmosphere, and that this increase has caused temperatures to rise. A few still disagree, and in March 2007 Channel 4 Television broadcast a documentary, *The Great Global Warming Swindle*, about these dissident opinions. One of the more puzzling pieces of evidence put forward in this programme was the observed long-term relation between temperature and CO_2. Former presidential candidate Al Gore, who has been very active trying to persuade the public that climate change is real, was shown

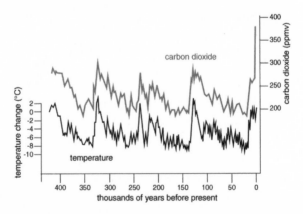

Historical records of temperature and CO_2, based on: J. R. Petit and others, 'Climate and atmospheric history of the past 420,000 years from the Vostok ice core, Antarctica', *Nature*, vol. 399, pp. 429–436 (1999).

delivering a lecture in front of a huge display of how temperature and CO_2 have changed in the past. These figures can be deduced from natural records such as ice cores.

The two curves go up and down almost together, a convincing association. But the programme pointed out that the temperature increases start and end *before* the CO_2 ones do, especially if you look closely at the most recent data. Clearly it is rising *temperature* that causes CO_2 to increase, not the other way round. This argument seems quite convincing, and the programme placed a lot of emphasis on it.

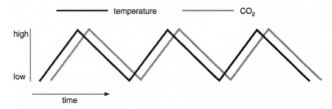

Temperature always changes first (schematic picture for illustrative purposes).

Climate science relies heavily on mathematical models of the physical processes that influence the climate, so this is a mathematical issue as well as a scientific one. The best data available to date indicate that this effect is real, with the CO_2 peaks and troughs appearing about 100 years after those of temperature. So does this relationship prove that rising temperatures cause rising CO_2, rather than the other way round? And what, if anything, does all this have to do with global warming?

Let's put our brains in gear first. The graphs are well known to climatologists, and indeed are a big part of the evidence that human production of CO_2 is causing temperatures to rise. If those graphs really do prove that CO_2 is *not* responsible for rising temperatures, the climatologists are likely to have noticed. Yes, it could all be a big conspiracy, but governments worldwide would be much happier if climate change turned out to be a delusion, and they're the ones paying for the research. If there's a conspiracy, it's far more likely to be one that tries to suppress evidence of climate change. So it seems likely that the climatologists have worked out why this delay occurs, and have concluded that it does not show that CO_2 plays no significant role in climate change. And indeed they have: it takes 30 seconds on the internet to find the explanation.

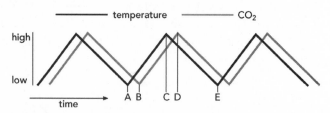

What happens at times A, B, C, D and E?

So why does that 100-year delay happen? The full story is complicated, but the broad outlines aren't hard to grasp if we think about the schematic picture, where the issues are easier to follow. The key facts are these:

- There is a natural cycle of changes in temperature caused by systematic changes in the Earth's orbit, the tilt of its axis, and the direction in which the axis points.
- Rises in temperature do indeed cause CO_2 levels to rise, with a time delay of tens or hundreds of years for nature to respond to the temperature change.

First, observe that most of the time, temperature and CO_2 levels rise together (between times B and C), or fall together (between times D and E). This shows temperature and CO_2 are linked, but it doesn't tell us which is cause and which is effect. In fact, each causes the other.

What is actually going on here, according to the vast majority of climatologists, is roughly this. At time A the natural cycle causes temperatures to start rising, though not by much. By time B, a century or so later, the effect on CO_2 becomes apparent. This rise feeds back into the temperature, which responds far more quickly to CO_2 levels than CO_2 levels do to temperature. So the temperature rises. Now temperature and CO_2 reinforce each other though positive feedback, and both climb together (times B to C). At time C, the external temperature cycle and other factors cause temperatures to begin to fall. The CO_2 levels don't show much effect until time D, but as soon as they do, the fall in CO_2 reinforces the fall in temperature, and both drop together. This continues until time E, when the whole process repeats itself.

Next question: what does this have to do with global warming?

Not a lot.

What we've been discussing is a natural free-running cycle, without human intervention. The terms 'global warming' and 'climate change' do not refer to increasing temperatures or changes in climate *as such*. They refer, very specifically, to deviations from the natural cycle.

The term 'global warming' was used first by scientists who understood that point and also understood that what was being discussed was medium-term average global temperatures, not short-term local ones. It was widely misunderstood, because some parts of the globe may cool for a time, while others get warmer. So the term 'climate change' started to be used in the hope of avoiding confusion. But that phrase doesn't just mean 'the climate is changing': that happens during the natural cycle. It means 'the climate is changing in a way that the natural cycle does not explain'.

In the natural cycle, as we have seen, temperature influences CO_2 and CO_2 influences temperature. When the atmosphere is 'forced' by a changing cycle of solar radiation, both quantities respond. The issue of 'global warming' is: what do we expect to happen to that cycle if humans cause large quantities of CO_2 to enter the atmosphere? Mathematically, this amounts to giving a big kick to CO_2, and seeing what the system does. And the answer is: the temperature promptly goes up too, because it responds fairly rapidly to changes in CO_2.

So the graphs, with that puzzling time delay, show what a

free-running atmospheric system does when it is forced by variations in incoming radiation. 'Global warming' isn't about that at all. It's about what this free-running system will do when you give it a kick. We know that human activity has raised CO_2 levels significantly over the past 50 years or so; in fact, they are now distinctly higher than anything found earlier in the ice core record. Look at the right-hand end of the graph of CO_2 on page 164. The proportions of various isotopes of carbon (different forms of the carbon atom with different atomic weight) show that this rise is mainly the result of human activity – and the unprecedented level of CO_2 in modern times confirms that.

To test the hypothesis that this rise in CO_2 has led to global warming, the mathematical kick that we give to any model of the atmosphere also has to be a rise in CO_2. So we are asking what effect this rise in CO_2 *causes* – in that context.

To check what happens, and to make it clear that this really is mathematics, I set up a simple system of model equations for how temperature T and carbon dioxide levels C change over time. It's not 'realistic', but it has the basic features we are discussing, and illustrates the key point. It looks like this:

$$\frac{dT}{dt} = \sin t + 0.25C - 0.01T^2$$

$$\frac{dC}{dt} = 0.1T - 0.01C^2$$

Here temperature is forced periodically (the $\sin t$ term) which models the changing heat coming from the Sun. Moreover, any change in C produces a proportionate change in T (the $0.25C$ term), and any change in T produces a proportionate change in C (the $0.1T$ term). So my model is set up so that higher temperatures cause more CO_2, and more CO_2 causes higher temperatures, just like the real world. Since 0.25 is bigger than 0.1, temperature responds faster to changes in CO_2 than CO_2 does to changes in temperature. Finally, I subtract $0.01T^2$ and $0.01C^2$ to mimic the cut-off effects known to occur.

I now solve these equations on my computer, and see what I get. Here are three pictures of how T (black curve) and C (grey curve) change over time. I have plotted $4y - 60$ rather than y to move the two curves close enough together to see the relationship.

- When the system is free-running, both T and C fluctuate periodically, and C *lags behind* T. This is the paradoxical time delay, which according to the TV programme means that rising CO_2 does not cause rising temperatures. But, in our model, rising CO_2 does cause rising temperatures, thanks to the $0.25C$ term in the first equation, yet we still see that time delay. The time delay is a consequence of non-linear effects in the model, not delays in what affects what.

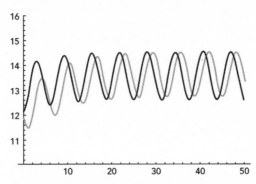

How temperature (black line) and CO_2 (grey line) vary over time. Note that CO_2 lags behind temperature.

- When I give C a sudden brief increase at time 25, both T and C react. However, C still appears to lag behind T, and T doesn't seem to change much.

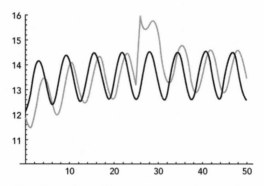

The effect of a sudden increase in CO_2 (grey line).

- However, if I graph the *changes* in T and C between the two runs of the equations, then I see that T starts to increase as soon as C does. So a change in C *does* cause an immediate change in T. What's interesting here is how temperature *continues* to increase while the spike in CO_2 is dying down. Non-linear dynamics can be counter-intuitive, which is why we have to use mathematics rather than naive verbal arguments.

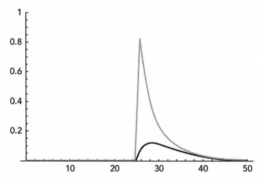

Differences in CO_2 and temperature between the two runs show that temperature rises immediately CO_2 does.

So the issue of 'global warming' or 'climate change' is not what causes what in the free-running system, where both rising temperature and rising CO_2 cause each other. Climate scientists don't dispute that and have known about it for a long time. The issue is: what happens when we *know* that one of these quantities has suddenly been changed by human activity? That much-trumpeted time delay is irrelevant to this question – in fact, it is misleading. The resulting temperature *change* begins immediately, and rises.

For further information, take a look at:
en.wikipedia.org/wiki/Climate_change
en.wikipedia.org/wiki/Global_warming

And you may find it informative to look at what happened *after* that Channel 4 broadcast, at:
en.wikipedia.org/wiki/The_Great_Global_Warming_Swindle

• •

Name the Cards

'Ladies and gentlemen,' the Great Whodunni announced, 'My assistant Grumpelina will ask a member of the audience to place three cards in a row on the table, while I am blindfolded. I will then ask her to provide some limited information, after which I will name those cards.'

The cards were chosen and placed in a row. Grumpelina then recited a strange list of statements:

'To the right of a King there's a Queen or two.

'To the left of a Queen there's a Queen or two.

'To the left of a Heart there's a Spade or two.

'To the right of a Spade there's a Spade or two.'

Instantly, Whodunni named the three cards.

What were they? [Note that 'two' here means two cards, not a card with two spots.]

Answer on page 317

• •

What Is Point Nine Recurring?

The first place most of us encounter mathematical infinity is when we study decimals. Not only do exotic numbers like π 'go on for ever' – so do more prosaic ones. Probably the first example we get to see is the fraction $\frac{1}{3}$. In decimals, this becomes $0.333333\ldots$, and the only way to make the decimal *exactly* equal to $\frac{1}{3}$ is to let it continue for ever.

The same problem arises for any fraction p/q where q is not just a lot of 2's and 5's multiplied together (which in particular includes all powers of 10). But unlike π, the decimal form of a fraction repeats the *same* pattern of digits over and over again, perhaps after some initial digits that don't fit that pattern. For instance, $\frac{83}{35} = 2.3714285714285714285\ldots$, repeating the 714285 indefinitely. These are called recurring decimals, and the part that repeats is usually marked with a dot, or dots at each end if it involves several digits:

$$\frac{1}{3} = 0.\dot{3}, \qquad \frac{83}{35} = 2.3\dot{7}1428\dot{5}$$

All this sounds reasonable, but the number $0.999999\ldots$, or $0.\dot{9}$, often causes trouble. On the one hand, it is obviously equal to 3 times $0.\dot{3}$, which is $3 \times \frac{1}{3}$, which is 1. On the other hand, 1 in decimals is $1.000000\ldots$, which doesn't look the same.

It seems to be widely believed that $0.\dot{9}$ is slightly less than 1. The reason for thinking that is presumably that whenever you stop, say at 0.9999999999, the resulting number differs from 1. The difference isn't very big – here it's 0.0000000001 – but it's not zero. But, of course, the point is that you shouldn't stop. So that argument doesn't hold water. Nevertheless, many people get a sneaky feeling that $0.\dot{9}$ *still* ought to be less than 1. How much less? Well, by a number that is smaller than anything looking like $0.000\ldots01$, no matter how many 0's there are.

A friend of mine, who worked in mathematics education, used to ask people how big $0.\dot{3}$ is, and then how big $0.\dot{9}$ is. Everyone was happy that the first decimal is exactly $\frac{1}{3}$, but on being told to multiply by 3, they became nervous. One said: 'That's sneaky! At first I thought that point three recurring is exactly one-third, but now I see it must be slightly *less* than one-third!'

We get confused about this point because it's a subtle feature of infinite series, and though we all do decimals, we don't do infinite series at school. To see the connection, observe that

$$0.\dot{9} = \frac{9}{10} + \frac{9}{100} + \frac{9}{1,000} + \frac{9}{10,000} + \cdots$$

This series *converges*, that is, it has a well-defined sum, and the rules of algebra apply. So we can use a standard trick. If the sum is s, then

$$10s = 9 + \frac{9}{10} + \frac{9}{100} + \frac{9}{1,000} + \cdots = 9 + s$$

so $9s = 9$, and $s = 1$.

There are lots of other calculations like this. They all tell us that $0.\dot{9} = 1$.

So what about that number that is smaller than anything looking like $0.000\ldots01$, no matter how many 0's there are? Is it an 'infinitesimal' – whatever that may mean?

Not in the real number system, no. There the only such number is 0. Why? Any (small) non-zero number has a decimal representation with a lot of 0's, but eventually some digit must be non-zero – otherwise the number is $0.000\ldots$, which is 0. As soon as we reach that position, we see that the number is greater than or equal to $0.000\ldots01$ with the appropriate number of 0's. So it doesn't satisfy the definition. In short: the difference between 1 and $0.\dot{9}$ is 0, so they are equal.

This is an annoying feature of the decimal representation: some numbers can be written in two apparently different ways. But the same goes for fractions: $\frac{1}{3}$ and $\frac{2}{6}$ are equal, for instance. No worries. You get used to it.

Ghost of a Departed Quantity

After decades of institutionalized denial, research mathematician reveals: .999... can be less than one, almost everywhere.

It took mathematicians centuries of effort to hammer out a logically rigorous theory of limits, infinite series and calculus, which they called 'analysis'. All the seductive but logically incoherent ideas about infinitely large and infinitely small numbers – infinitesimals – were safely banished. The philosopher George Berkeley had scathingly referred to infinitesimals as 'ghosts of departed quantities', and everyone agreed he was right. However, calculus worked anyway, thanks to limits, which exorcised the ghosts.

Infinity, big or small, was a process, not a number. You never added all the terms of an infinite series: you added a finite number, and asked how that sum behaved as the number of terms grew ever larger. You approached infinity, but you never *got* there. Similarly, infinitesimals don't exist. No positive number can be smaller than *any* positive number, because then it has to be smaller than itself.

But, as I say somewhere else in this book, you should never give up on a good idea just because it doesn't work. Around 1960, Abraham Robinson made some surprising discoveries at the frontiers of mathematical logic, reported in his 1966 book *Non-Standard Analysis*. He proved that there are extensions of the real number system (called 'non-standard reals' or 'hyperreals') that share almost all the usual properties of real numbers, except that infinite numbers and infinitesimals genuinely do exist. If n is an infinite number, then $1/n$ is infinitesimal – but not zero. Robinson showed that the whole of analysis can be set up for hyperreals, so that, for example, an infinite series is the sum of infinitely many terms, and you *do* get to infinity.

An infinitesimal is now a new kind of number that is smaller than any positive real number, but it is not *itself* a real number. And it is not smaller than any positive hyperreal number. But you can convert all finite hyperreals to real numbers by taking the 'standard part', which is the unique real number that is infinitesimally close.

There is a price to pay for all this. The proof that hyperreals exist is non-constructive – it shows they can occur, but doesn't tell you what they are. However, any theorem about ordinary analysis that can be proved using non-standard analysis has some standard-analysis proof. So this is a new method for proving the same theorems about ordinary analysis, and it is closer to the intuition of people like Newton and Leibniz than the more technical methods introduced later.

There have been some attempts to introduce non-standard analysis into undergraduate mathematics teaching, but the approach remains a minority sport. For more information, go to: en.wikipedia.org/wiki/Non-standard_analysis

As I was writing this book, and had just finished the previous item on $0.\dot{9}$, Mikhail Katz emailed me a paper, written with Karin Usadi Katz, that uses non-standard analysis to place that expression in a different light. They point out that in ordinary analysis there is an exact formula

$$\frac{9}{10} + \frac{9}{10^2} + \frac{9}{10^3} + \cdots + \frac{9}{10^n} = 1 - \left(\frac{1}{10}\right)^n$$

for any finite decimal $0.999\ldots9$. Now let n be an infinite hyperreal. The same formula holds, but when n is infinite, $\left(\frac{1}{10}\right)^n$ is *not* zero, but infinitesimal. The departed quantity does indeed leave behind a ghost.

Similar remarks hold for the infinite series that represents $0.\dot{3}$. None of this contradicts what I said earlier about $0.\dot{9}$ and $0.\dot{3}$, because then I was talking about standard analysis, and the standard part of $1 - \left(\frac{1}{10}\right)^n$ is 1 when n is infinite. But it shows that the intuitive feeling some people have, that 'there's a little bit missing', can be given a rigorous justification if it is interpreted in an entirely reasonable way. I don't think we should teach that approach in school, but it should make us more sympathetic to anyone who suffers from that particular difficulty.

Katz and Katz's paper contains a lot more about this issue,

and poses the key question: 'What does the teacher mean to happen exactly after *nine, nine, nine* when he writes *dot, dot, dot*?' The standard analysis answer is to take ' . . . ' as indicating passage to a limit. But in non-standard analysis there are many different interpretations. The traditional one assigns the largest possible sensible value to the expression – which is 1. But there are others.

Nice Little Earner

Smith and Jones were hired at the same time by Stainsbury's Superdupermarket, with a starting salary of £10,000 per year. Every six months, Smith's pay rose by £500 compared with that for the previous 6-month period. Every year, Jones's pay rose by £1,600 compared with that for the previous 12-month period. Three years later, who had earned more?

Answer on page 317

A Puzzle for Leonardo

In 1225, Emperor Frederick II visited Pisa, where the great mathematician Leonardo (later nicknamed Fibonacci; see *Cabinet*, page 98) lived. Frederick had heard of Leonardo's reputation, and – as emperors do – he thought it would be a great idea to set up a mathematical tournament. So the emperor's team, which consisted of John of Palermo and Theodore, but not the emperor, battled it out head-to-head with Leonardo's team, which consisted of Leonardo.

Among the questions that the emperor's team set Leonardo was this: find a perfect square which remains a perfect square when 5 is added or subtracted. They wanted a solution in rational numbers – that is, fractions formed by whole numbers.

Help Leonardo solve the emperor's puzzle.

Answer on page 318. Or see the next item.

Congruent Numbers

Emperor Frederick II's question in the previous puzzle[*] leads into deep mathematical waters, and only recently have mathematicians begun to plumb their murky depths. The question is: what happens if we replace 5 by an arbitrary whole number? For which whole numbers d can we solve

$$y^2 - d = x^2, \qquad y^2 + d = z^2$$

in rational numbers x, y, z?

Leonardo called such d 'congruent numbers', a term still used today despite it being a bit confusing – number theorists habitually use the word 'congruent' in a completely different way. Congruent numbers can be characterised as the areas of rational Pythagorean triangles – right-angled triangles with rational sides. This isn't obvious, but it's true: Leonardo's method of solution, explained in the answer to the previous problem, hints at this result. If the triangle has sides a, b, c with $a^2 + b^2 = c^2$, then its area is $ab/2$. Let $y = c/2$. Then a calculation shows that $y^2 - ab/2$ and $y^2 + ab/2$ are both perfect squares. Conversely, we can construct a Pythagorean triangle from any solution x, y, z, d, with d equal to the area.

The familiar 3-4-5 triangle has area $3 \times 4/2 = 6$, so 6 is a congruent number. Here the recipe tells us to take $y = 5/2$. Then

$$x^2 = \frac{25}{4} - 6 = \frac{1}{4}, \qquad \text{so } x = \frac{1}{2}$$

---

[*] It was probably suggested by John of Palermo, but it is the emperor's question all the same, just as the Great Pyramid was indisputably built by the pharaoh Khufu. Emperors are like that. Hans Christian Anderson's story of the emperor's new clothes is completely unconvincing: any little boy who dared to contradict the emperor would have ended up in jail. The cliché 'the emperor has no clothes' affirms the imperial status – what people usually mean is that the clothes contain no emperor, which isn't quite the same thing.

$$z^2 = \frac{25}{4} + 6 = \frac{49}{4}, \quad \text{so } z = \frac{7}{2}$$

To get $d = 5$, we have to start with the 40-9-41 triangle, with area $180 = 5 \times 36$. Then divide by $6^2 = 36$ to get the triangle with sides 20/3, 3/2, 41/6, whose area is 5. Now

$$x^2 = \frac{1681}{144} - 5 = \frac{961}{144}, \qquad \text{so } x = \frac{31}{12}$$

$$z^2 = \frac{1681}{144} + 5 = \frac{2401}{144}, \qquad \text{so } z = \frac{49}{12}$$

and we have recovered Leonardo's answer to the emperor's question.

The question now remains: which whole numbers d can be the area of a Pythagorean triangle with rational sides? The answer is not obvious. It turns out to be linked to a different equation,

$$p^2 = q^3 - d^2 q$$

which has solutions p, q in whole numbers if and only if d is congruent.

Some numbers are congruent, some aren't. For example, 5, 6, 7 are congruent, but 1, 2, 3, 4 aren't. It need not be straightforward to decide: for example, 157 is a congruent number, but the simplest right triangle with area 157 has hypotenuse

$$c = \frac{224403517704336969924557513090667486316094847204\,1}{8912332268928859588025535178967163570016480830}$$

The best test currently known depends on an unproved conjecture, the Birch–Swinnerton-Dyer conjecture, which is one of the Clay millennium mathematics prizes (*Cabinet*, page 127) with a million dollars on offer for a proof or disproof. Frederick II didn't realise what he was starting.

Present-Minded Somewhere Else

Norbert Wiener pioneered the mathematics of random processes, as well as the new subject of cybernetics, in the first half of the 20th century. He was a brilliant mathematician, and notorious for forgetting things. So when the family moved to a new house, his wife wrote the address on a slip of paper and gave it to him. 'Don't be silly, I'm not going to forget anything as important as *that*,' he said, but he put the paper in his pocket anyway.

Norbert Wiener.

Later that day, Wiener became immersed in a mathematical problem, needed some paper to write on, took out the slip bearing his new address, and covered it in equations. When he had finished these rough calculations, he crumpled the paper into a ball and threw it away.

As evening approached, he recalled something about a new house but couldn't find the slip of paper with its address. Unable to think of anything else to do, he walked to his old house, and noticed a little girl sitting outside it.

'Pardon me, my dear, but do you happen to know where the Wieners have mov—'

'That's OK, Daddy. Mommy sent me to fetch you.'

It's About Time

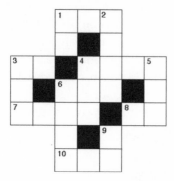

Crossnumber grid.

A crossnumber is like a crossword, but using numbers instead of words. All the clues for this one are about time, and are prefaced by the phrase 'the number of ...'.

Across
1 Days in a normal year
3 Minutes in a quarter of an hour
4 Seconds in one hour, 24 minutes and 3 seconds
6 Seconds in five minutes
7 Hours in a normal year
8 Hours in 4 days
10 Days in a leap year

Down
1 Days in October
2 Seconds in an hour and a half
3 Hours in a week
4 Hours in 20 days 20 hours
5 Hours in a fortnight
6 Seconds in one hour 3 seconds
9 Hours in a day and a half

Answer on page 319

Do I Avoid Kangaroos?

- The only animals in this house are cats.
- Every animal that loves to gaze at the moon is suitable for a pet.
- When I detest an animal, I avoid it.
- No animals are meat-eaters, unless they prowl by night.
- No cat fails to kill mice.

- No animals ever take to me, except those in this house.
- Kangaroos are not suitable for pets.
- Only meat-eaters kill mice.
- I detest animals that do not take to me.
- Animals that prowl at night love to gaze at the moon.

If all these statements are correct, do I avoid kangaroos, or not?

Answer on page 319

The Klein Bottle

In the late 1800s, there was a vogue for naming special surfaces after mathematicians: Kummer's surface, for instance, was named after Ernst Eduard Kummer. The mathematicians tended to be German, and the German word for surface is *Fläche*, so this was the 'Kummersche Fläche'. I'm delving into the linguistics here, because it led to a pun being used to name a mathematical concept. That still happens, but this may well have been the first occasion. The pun derives from a very similar word, *Flasche*, which means 'bottle'. At any rate, the scene was set: when Felix Klein invented a bottle-shaped surface in 1882, it was naturally called the 'Kleinsche Fläche'. And inevitably this rapidly mutated into 'Kleinsche Flasche' – the Klein bottle.

I don't know whether the pun was deliberate, or a mistranslation. At any rate, the new name was so successful that even the Germans adopted it.

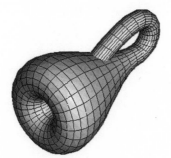

Klein's surface . . .

... interpreted as a bottle.

The Klein bottle is important in topology, as an example of a surface with no edges and only one side. A conventional surface, such as the sphere – by which topologists just mean the thin skin of the sphere's surface, and not a solid ball (which they call a ball) – has two distinct sides, an inside and an outside. You can imagine painting the inside red and the outside blue, and the two colours never meet. But you can't do that with a Klein bottle. If you start painting what looks like the outside blue, you get to the bent tube where it becomes narrower, and if you follow that tube as it penetrates through the bulging body you end up painting what looks like the inside blue as well.

Klein invented his bottle for a reason: it came up naturally in the theory of Riemann surfaces in complex analysis, which classifies nasty kinds of behaviour – in a beautiful way – when you try to develop calculus over the complex numbers. The Klein bottle is reminiscent of an even more famous surface, the Möbius band (or strip), formed by twisting a strip of paper and gluing the ends together. The Möbius band has one side, but it also has an edge (*Cabinet*, page 111). The Klein bottle gets rid of the edge, which topologists find more convenient because edges can cause trouble. Especially in complex analysis.

There is a price to pay, however: the Klein bottle can't be represented in ordinary 3D space without penetrating through itself. However, topologists don't mind that, because they don't represent their surfaces in 3D space anyway. They prefer to think of them as abstract forms in their own right, not relying on the existence of a surrounding space. In fact, you can fit a Klein bottle into 4D space without any interpenetration, but that brings its own difficulties.

One way to represent a Klein bottle, which doesn't require any self-intersection, is to borrow a trick that is familiar to almost everyone nowadays from computer games. (The topologists thought of it long before, I hasten to add.) In many games, the flat rectangular computer screen is 'wrapped round' so that the left and right edges are in effect joined together. If an alien spaceship shoots off the right-hand edge, it immediately reappears at the left-hand edge. The top and bottom may also be wrapped round in this manner. Now, a computer screen doesn't actually *bend*. Much. So the 'wrapping round' is purely conceptual, a figment of the programmer's mind. But we can easily *imagine* that the opposite edges abut, work out what would happen if they did, and respond accordingly. And that's what topologists do.

Specifically, they also start with a rectangle, and wrap its edges round so that in the imagination they join. But there's a twist – literally. The top and bottom edges are wrapped round as usual, but the right edge is given a half-twist, interchanging top and bottom, before wrapping it round to meet the left edge. So when a spaceship shoots off the top, it reappears from the corresponding position at the bottom; but when it shoots off the right-hand edge, it reappears upside down and at the opposite end of the left-hand edge.

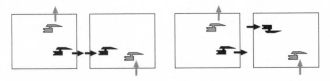

Conventional computer screen   Klein bottle wrap-round.
wrap-round.

Topologically, the conventional wrapped-round screen is a torus – like a car inner tube or (I have to say this because many people have never *seen* an inner tube, since most car tyres are tubeless) a doughnut. But only the sugary surface, not the actual dough. You can see why if you imagine what happens when you do actually join the edges – using a flexible screen. Joining top to bottom creates a cylindrical tube; then joining the ends of the tube bends it round into a closed loop.

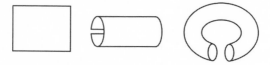

Wrapping round without a twist creates a torus.

However, if you imagine a similar procedure for a Klein bottle, then the two ends of the cylinder don't join up that way: one of them has to be given the opposite orientation. In 3D, one way to do this is to make it thinner, poke it through the side of the cylinder, poke it out of the open end, and then roll it back on itself like the neck of a sweater and finally join it to the other end of the cylinder. This leads to the standard 'bottle' shape, with a self-intersection where you poked it through. As Klein wrote: the shape 'can be visualized by inverting a piece of a rubber tube and letting it pass through itself so that outside and inside meet'.

Joining the edges of a cylinder to make a Klein bottle.

With an extra dimension to play with, you can push the end of the cylinder off into the fourth dimension before poking it through where the cylinder would have been; then pull it back into normal 3D space once it's inside, and carry on as normal. That way, there's no self-intersection.

The Klein bottle has a remarkable property, which has been celebrated in a limerick, whose author – perhaps mercifully – remains unknown:

> A mathematician named Klein
> Thought the Möbius band was divine.
> Said he: 'If you glue
> The edges of two,
> You'll get a weird bottle like mine.'

Can you see how to achieve this?

Answer on page 320

For some brilliant visualisations, go to:
plus.maths.org/issue26/features/mathart/index-gifd.html

Another cute factoid: any map on the Klein bottle can be coloured with at most 6 colours, so that adjacent regions have different colours. This compares with 4 colours for the sphere or plane (*Cabinet*, page 10) and 7 for the torus. See:
mathworld.wolfram.com/KleinBottle.html

• •

Accounting the Digits

In the Great Celestial Number Factory, where all numbers are made, the accountants keep tabs on how many times each digit 0–9 is used, to make sure that there are adequate stocks in the warehouse. They record these counts on a standard form, like this:

Typical inventory form.

So, for example, as the digit 4 occurs 3 times, Nugent writes '3' in the lower row of boxes, underneath the printed 4. Numbers are written so that they end in the right-hand box, like the example, and leading zeros may or may not occur. (None of that matters for this puzzle, but people do *worry*...)

One day Nugent was filling in the form, as usual, when he suddenly noticed something remarkable: the two numbers (that is, sequences of digits) recorded in the two rows of boxes were *identical*.

What was the number concerned?

Answer on page 321

• •

Multiplying with Sticks

We all know how to measure a length when our ruler or tape measure is too short. We measure as far as we can, mark the end point, then continue measuring from there, and add the distances together. This puts into practice a basic principle of Euclid's geometry: if you place two lines end to end – pointing in the same direction – then their lengths add.

This means that you can make an adding machine from two sticks. Just make marks along the edge distances 1, 2, 3, 4, and so on; then position the sticks to perform the addition sum.

The number on the top stick is 3 more than the corresponding one on the bottom stick.

Big deal, I hear you thinking, and it's true that this gadget isn't terribly practical. But a close relative is – or, to be honest, was. To get it, we change the markings, replacing each number by the corresponding power of 2.

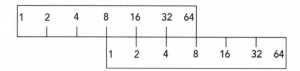

Now the numbers on the top stick are the corresponding numbers on the bottom stick, *multiplied by 8*. Our adding-sticks have become multiplying-sticks. This trick works because of the well-known formula

$$2^a \times 2^b = 2^{a+b}$$

Well, that's fantastic. Now we can multiply powers of 2.

Back in the days when computers and calculators were undreamt of, and would have been seen as magic, multiplying two numbers was really hard work. But astronomers need to do a lot of multiplications to keep track of the stars and planets. So, around 1594, James Craig, court doctor to King James VI of Scotland, told John Napier, Baron of Murchiston, about something called *prosthaphaeiresis*. It sounds painful, and in a way it was: the Danish mathematicians had discovered how to multiply numbers using a formula discovered by François Viète:

$$\sin\frac{x+y}{2}\cos\frac{x-y}{2} = \frac{\sin x + \sin y}{2}$$

Using tables of sines and cosines, you could use this formula to turn a multiplication problem into a short series of addition problems. It was a bit complicated, but it was still quicker than conventional multiplication methods.

For years Napier had been thinking about efficient methods for doing sums, and it dawned on him that there was a better way. The formula for multiplying powers of 2 works for powers of any fixed number. That is,

$$n^a \times n^b = n^{a+b}$$

for any number n. If you set n to something close to 1, such as 1.001, then the successive powers will be very closely spaced, so any number that interests you will be close to some power of n. Now you can use the formula to convert multiplication to addition. For instance, suppose I want to multiply 3.52 by 7.85. Well, to a good approximation

$$(1.001)^{1259} = 3.52$$
$$(1.001)^{2062} = 7.85$$

Therefore,

$$3.52 \times 7.85 = (1.001)^{1259} \times (1.001)^{2062} = (1.001)^{1259+2062}$$
$$= (1.001)^{3321} = 27.64$$

The exact answer is 27.632. Not bad!

Pages from Napier's logarithm tables.

For more accuracy, you should replace 1.001 by something more like 1.0000001. Then you just draw up a table of the first million or so powers of that number, and you've got a quick way to multiply numbers to about 9-digit accuracy, just by adding the corresponding powers. Perversely, Napier chose to use powers of 0.9999999, which is less than 1, so the numbers got smaller as the powers got larger.

Fortunately, Henry Briggs, an Oxford professor, took an interest and sorted out a better way. The upshot of all this was the concept of a logarithm, which turns the calculations back to front. For example, since $(1.001)^{1259} = 3.52$, the logarithm of 3.52 to base 1.001 is 1,259. In general, $\log x$ (to base n) is whichever number a satisfies

$$n^a = x$$

Now the formula for n^{a+b} can be reinterpreted as

$$\log xy = \log x + \log y$$

whichever base you use. For practical purposes, base 10 is best, because we use decimals. Mathematicians prefer base e, which is roughly 2.71828, because it is better behaved with respect to the operations of calculus.

All very well, but what does this have to do with sticks? Well, what we're doing, in effect, with those powers of 2, is marking each number at a distance along the stick given by its logarithm. For example, since $2^5 = 32$, the logarithm of 32 to base 2 is 5, so we write 32 five units along the stick.

We have now invented the slide rule, which is basically a table of logarithms written in wood. We were anticipated around 1600 by William Oughtred and others, who over the centuries added many more scales for trigonometric functions, powers, and other mathematical operations. The slide rule – colloquially called a *slipstick** – was widely used by scientists and (especially) engineers until about 40 years ago, when it was rendered obsolete by electronic calculators.

* You can give a slipstick to a pig, but it's still a pig.

A slide rule from the sixties.

Today the slide rule is mostly a quaint reminder of the predigital age. I own two: one I used at school, mainly in physics lessons, and a bamboo one I bought in a flea market. To find out more, visit:

en.wikipedia.org/wiki/Slide_rule
www.sliderule.ca/
www.sliderules.info/

• •

As Long as I Gaze on Laplacian Sunrise

Pierre Simon de Laplace is best known for his work in celestial mechanics, but he was also one of the pioneers of probability theory. Now, pioneering work is often sloppy, because the basic issues haven't been properly explored; that's what pioneers are for, in fact.

Laplace argued that, if we observe the Sun rising every morning for $n - 1$ days, then we can infer that the probability that it will *not* rise the next morning is $1/n$. After all, out of n mornings, it has risen on $n - 1$, so only 1 is left for it not to rise.

Ignoring the dodgy assumptions here, there is a reassuring deduction: since the Sun has now risen for hundreds of billions of consecutive mornings, the probability that it won't rise tomorrow is staggeringly small.

Unfortunately, Laplace's argument has a sting in the tail. Accepting his value for the successive probabilities, what is the probability that the Sun will always rise?

Answer on page 321

• •

Another Take on Mathematical Cats

- Did Erwin Schrödinger have a cat?
  Yes and no.
- Did Werner Heisenberg have a cat?
  I'm not sure.
- Did Kurt Gödel have a cat?
  If he did, we can't prove it.
- Did Fibonacci have a cat?
  He certainly had a lot of rabbits.
- Did René Descartes have a cat?
  He thought he did.
- Did Augustin-Louis Cauchy have a cat?
  That's a complex question.
- Did Georg Bernhard Riemann have a cat?
  That hypothesis has not yet been proved.
- Did Albert Einstein have a cat?
  One of his relatives did.
- Did Luitzen Brouwer have a cat?
  Well, he didn't not have one.
- Did William Feller have a cat?
  Probably.
- Did Ronald Aylmer Fisher have a cat?
  The null hypothesis is rejected at the 95% level.

Bordered Prime Magic Square

Recall that a magic square is a square array of numbers, such that all rows, columns and diagonals have the same sum.

2777	1409	2339	1481	1061	2699	2087
2531	1889	2237	2459	1229	2081	1427
1367	2357	2399	1511	2027	1601	2591
2909	1031	1607	1979	2351	2927	1049
1301	2741	1931	2447	1559	1217	2657
1097	1877	1721	1499	2729	2069	2861
1871	2549	1619	2477	2897	1259	1181

Bordered prime magic square.

Allan Johnson, Jr, discovered a 7×7 magic square composed entirely of primes. Moreover, it is *bordered*: that is, the smaller 5×5 and 3×3 squares indicated by the bold lines in the picture are also magic.

• •

The Green–Tao Theorem

An arithmetic sequence* is a list of numbers such that successive differences are all equal – for example,

$$17, \quad 29, \quad 41, \quad 53, \quad 65, \quad 77, \quad 89$$

where each number is 12 greater than the one before. This is called the *common difference*.

In this particular list, which has seven terms, many numbers are prime, but some (65 and 77) aren't. However, it is possible to find seven primes in arithmetic sequence:

---

* Stress the third syllable, arith*me*tic, not a*rith*metic. An older term is 'arithmetic progression'.

$$7, \quad 37, \quad 67, \quad 97, \quad 127, \quad 157$$

with common difference 30.

Until recently, very little was known about the possible lengths of prime arithmetic sequences. There are infinitely many of length 2, because any two primes form an arithmetic sequence (there is only one difference, which equals itself) and there are infinitely many primes. In 1933 Johannes van der Corput proved that there are infinitely many prime arithmetic sequences of length 3, and there the matter rested.

Experiments, using computers when the numbers get big, found examples of prime arithmetic sequences with any length up to (as I write) 25. Here's a table:

Length k	Prime arithmetic sequence ($0 \le n \le k - 1$)
3	$3 + 2n$
4	$5 + 6n$
5	$5 + 6n$
6	$7 + 30n$
7	$7 + 150n$
8	$199 + 210n$
9	$199 + 210n$
10	$199 + 210n$
11	$110{,}437 + 13{,}860n$
12	$110{,}437 + 13{,}860n$
13	$4{,}943 + 60{,}060n$
14	$31{,}385{,}539 + 420{,}420n$
15	$115{,}453{,}391 + 41{,}44{,}140n$
16	$53{,}297{,}929 + 9{,}699{,}690n$
17	$3{,}430{,}751{,}869 + 8{,}729{,}721n$
18	$4{,}808{,}316{,}343 + 717{,}777{,}060n$
19	$8{,}297{,}644{,}387 + 4{,}180{,}566{,}390n$
20	$214{,}861{,}583{,}621 + 18{,}846{,}497{,}670n$
21	$5{,}749{,}146{,}449{,}311 + 26{,}004{,}868{,}890n$
22	$1{,}351{,}906{,}725{,}737{,}537{,}399 + 13{,}082{,}761{,}331{,}670{,}030n$
23	$117{,}075{,}039{,}027{,}693{,}563 + 1{,}460{,}812{,}112{,}760n$
24	$468{,}395{,}662{,}504{,}823 + 45{,}872{,}132{,}836{,}530n$
25	$6{,}171{,}054{,}912{,}832{,}631 + 81{,}737{,}658{,}082{,}080n$

There are others, but these have the smallest final term for given k.

In 2004, to general astonishment, the whole topic was blown out of the water by Ben Green and Terence Tao, who proved that there exist arbitrarily long prime arithmetic sequences. Their proof combined half a dozen different areas of mathematics, and it even gave an estimate of how small the primes could be, for a given k. Namely, they need be no larger than

$$2\char`^2\char`^2\char`^2\char`^2\char`^2\char`^2\char`^2\char`^100k$$

where $a\char`^b$ represents a^b. These numbers are mind-bogglingly large, and it is conjectured that they are much larger than necessary, and can be replaced by $k! + 1$. Here $k! = k \times (k-1) \times (k-2) \times \cdots \times 3 \times 2 \times 1$ is the factorial of k.

This theorem has many consequences. It implies that there exist arbitrarily large magic squares in which every row and every column consist of primes in arithmetic sequence. Indeed, the same goes for magic d-dimensional hypercubes, for any d.

In 1990, before Green and Tao proved their theorem, Antal Balog proved that, if that result were correct, then there would exist arbitrarily large sets of primes with the curious feature that the average of any two of them is also prime – and all these averages are different. For example, the six primes

$$3, \quad 11, \quad 23, \quad 71, \quad 191, \quad 443$$

form such a set, with all 15 averages (such as $(3+11)/2 = 7$ and $(23+443)/2 = 233$) being distinct primes. So now Balog's result is proved as well.

In the opposite direction, it has been known for a long time that every prime arithmetic sequence has finite length. That is, if you continue any arithmetic sequence for long enough you will hit a number that is not prime. This doesn't contradict the Green–Tao Theorem, because some *other* arithmetic sequence could contain more primes. So all lengths here are finite, but there is no upper limit to their sizes.

• •

Peaucellier's Linkage

In the early days of steam engines, there was a lot of interest in mechanical linkages that could turn rotary motion into straight-line motion, such as a wheel driving a pump. One of the neatest arrangements, which is mathematically exact, is Peaucellier's linkage, invented in 1864 by the French army officer Charles-Nicolas Peaucellier. It was invented independently by a Lithuanian named Lippman Lipkin.

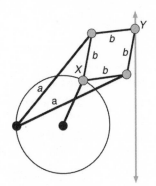

Peaucellier's linkage.

The two black blobs are fixed pins that let the links rotate; the grey ones are pins that link the rods together, also allowing them to rotate. The two rods marked a have the same length, and the four rods marked b have the same length. As pin X moves round the circle – which it must do because one rod is fixed to the centre of the circle – pin Y moves up and down along the straight line drawn in grey. The linkage limits the position of X to an arc of the circle, so Y is limited to a segment of the line.

The (fairly complicated) proof that it works, an animation of the linkage, and an explanation of the deeper mathematical ideas behind it can be found at:

en.wikipedia.org/wiki/Peaucellier-Lipkin_linkage

A Better Approximation to π

The famous approximation to π is 22/7, which is convenient for school calculations because it's nice and simple. It is *not* exact – in decimals,

$$22/7 = 3.142857142857\ldots$$

whereas

$$\pi = 3.141592653589\ldots$$

A more accurate approximation is

$$355/113 = 3.141592920353\ldots$$

which agrees with π to six decimal places – not bad for such a simple fraction. In fact, there is a rigorous sense in which 355/113 is the best approximation to π using numbers of that size.

The decimal for 22/7 keeps repeating the same sequence of digits, 142857, indefinitely. As mentioned on page 172, this is a general feature of fractions: if you write a fraction as a decimal, then either it stops, or it 'recurs': it goes on for ever, repeating the same string of digits over and over again. Conversely, all decimals that stop or recur are equal to exact fractions.

An example of a fraction whose decimal representation stops is

$$3/8 = 0.375$$

and one that repeats over and over again is

$$5/12 = 0.4166666\ldots$$

In a sense, the decimals for 3/8 *also* repeat for ever, because we can write

$$3/8 = 0.37500000000\ldots$$

with a repeating string 0. But terminating zeros are usually omitted.

It may not look as though the decimal for 355/113 repeats,

but actually it does – after the 112th decimal place! It is no coincidence that $112 = 113 - 1$, but it would take too long to explain why. If you take the calculation that far, you'll get

$$355/113 = 3.14159292035398230088495575221238938$$
$$05309734513274336283185840707964601 7$$
$$6991150442477876106194690265486725 66$$
$$37168\ldots$$

after which the digits repeat again, starting from immediately after the decimal point.

Because π is irrational – not equal to an exact fraction – its decimal expansion never repeats the same block of digits over and over again. This was proved in 1770 by Johann Lambert.

The next two approximations to π are 103,993/33,102 and 104,348/33,215.

• •

Strictly for Calculus Buffs

In 1944, D. P. Dalzell published a short note containing the curious formula

$$\int_0^1 \frac{x^4(1-x)^4}{1+x^2}dx = \frac{22}{7} - \pi$$

which relates π and its commonest approximation, 22/7, to an integral. You can verify the formula using no more than school calculus, because

$$\frac{x^4(1-x)^4}{1+x^2} = x^6 - 4x^5 + 5x^4 - 4x^2 + 4 - \frac{4}{1+x^2}$$

where the integral of each term is a standard result. The last term gives π and the rest give 22/7. This particular formula is significant, though, because the function being integrated is positive in the range from 0 to 1. The integral from 0 to 1 is just the average value, so this must also be positive. Since the function concerned is not always zero, we deduce that π is less than 22/7. This is a fairly simple way to prove that the usual approximation is not exact.

The formula also leads to an estimate of the error, because the maximum value of $x^4(1 - x^4)/(1 + x^2)$ between 0 and 1 is 1/256, so the average is at most 1/256. Therefore

$$\frac{5{,}625}{1{,}792} = \frac{22}{7} - \frac{1}{256} < \pi < \frac{22}{7}$$

With more effort you can prove that the error is at most 1/630.

This formula turns out to be part of a more extensive story (see page 322 for references). In 2005, Stephen Lucas started thinking about the improved approximation to π, 355/113, which we've just encountered. Lucas found the formula

$$\int_0^1 \frac{x^8(1 - x)^8(25 + 816x^2)}{3{,}164(1 + x^2)} dx = \frac{355}{113} - \pi$$

which in the circumstances is quite elegant. Again the function being integrated is positive, so the formula proves that π is (slightly) smaller than 355/113.

• •

The Statue of Pallas Athene

According to a puzzle book published in the Middle Ages, the statue of the goddess Pallas Athene was inscribed with the following information:

'I, Pallas, am made from the purest gold, donated by five generous poets. Kariseus gave half; Thespian an eighth. Solon gave one-tenth; Themison gave one-twentieth. And the remaining nine talents' worth of gold was provided by the good Aristodokos.'

How much did the statue cost in total? [A *talent* is a unit of weight, roughly 1 kilogram.]

Answer on page 322

How much gold?

●●

Calculator Curiosity 3

Get your calculator, and work out:

$$6 \times 6$$
$$66 \times 66$$
$$666 \times 666$$
$$6{,}666 \times 6{,}666$$
$$66{,}666 \times 66{,}666$$
$$666{,}666 \times 666{,}666$$
$$6{,}666{,}666 \times 6{,}666{,}666$$
$$66{,}666{,}666 \times 66{,}666{,}666$$

At least, do that until your calculator runs out of digits. After which you should be able to guess what happens anyway.

Answer on page 322

●●

Completing the Square

The traditional 3×3 magic square looks like this.

The traditional magic square.

Each cell contains a different number, and each row, column and diagonal sums to 15.

Your task is to find a square satisfying the same conditions, but with an 8 at top centre, like this:

	8	

Start here!

Answer on page 322

. . . .

The Look and Say Sequence

One of the strangest sequences in mathematics was invented by John Horton Conway. It begins

1 11 21 1211 111221 312211 13112221 1113213211

- What is the rule for forming the sequence? The title of this section is a hint.
- Roughly how long is the *n*th term in this sequence? [For experts only]

Answers on page 323

. . . .

Non-Mathematicians Musing About Mathematics

The things of this world cannot be made known without a knowledge of mathematics.   *Roger Bacon*

I had a feeling once about Mathematics – that I saw it all ...
I saw – as one might see the transit of Venus or even the Lord Mayor's Show – a quantity passing through infinity and changing its sign from plus to minus. I saw exactly why it happened and why the tergiversation was inevitable, but it was after dinner and I let it go.   *Sir Winston Spencer Churchill*

Mathematics seems to endow one with something like a new sense.   *Charles Darwin*

For a physicist, mathematics is not just a tool by means of which phenomena can be calculated; it is the main source of concepts and principles by means of which new theories can be created.
Freeman Dyson

Do not worry about your difficulties in Mathematics. I can assure you mine are still greater.   *Albert Einstein*

Equations are just the boring part of mathematics. I attempt to see things in terms of geometry.   *Stephen Hawking*

Anyone who cannot cope with mathematics is not fully human. At best he is a tolerable subhuman who has learned to wear shoes, bathe, and not make messes in the house.
Robert A. Heinlein

Mathematics may be compared to a mill of exquisite workmanship, which grinds your stuff to any degree of fineness; but, nevertheless, what you get out depends on what you put in; and as the grandest mill in the world will not extract wheat flour from peascods, so pages of formulae will not get a definite result out of loose data.   *Thomas Henry Huxley*

Medicine makes people ill, mathematics make them sad, and theology makes them sinful.   *Martin Luther*

I tell them that, if they will occupy themselves with the study of mathematics, they will find in it the best remedy against the lusts of the flesh.   *Thomas Mann*

The greatest unsolved theorem in mathematics is why some people are better at it than others.   *Adrian Mathesis*[*]

She knew only that if she did or said thus-and-so, men would unerringly respond with the complimentary thus-and-so. It was like a mathematical formula and no more difficult, for mathematics was the one subject that had come easy to Scarlett in her schooldays.   *Margaret Mitchell*

The advancement and perfection of mathematics are intimately connected with the prosperity of the State.   *Napoleon I*

Mathematical propositions express no thoughts ... we use mathematical propositions only in order to infer from propositions which do not belong to mathematics to others which equally do not belong to mathematics.
Ludwig Wittgenstein

[Mathematics] is an independent world.
Created out of pure intelligence.   *William Wordsworth*

I'm sorry to say that the subject I most disliked was mathematics. I have thought about it. I think the reason was that mathematics leaves no room for argument. If you made a mistake, that was all there was to it.   *Malcolm X*

Like the crest of a peacock, so is mathematics at the head of all knowledge.   *An old Indian saying*

• •

[*]   Who appears to be pseudonymous.

Euler's Conjecture

Fermat's Last Theorem states that two non-zero integer cubes can't add up to a cube, and ditto for fourth, fifth or higher powers. It was famously proved by Andrew Wiles in 1994–5 (*Cabinet*, page 50). One of the first people to make inroads into the problem was Euler, who proved the Last Theorem for cubes: two non-zero cubes cannot add up to a cube. But he also noticed that three cubes can add up to a cube. In fact,

$$3^3 + 4^3 + 5^3 = 6^3$$

Euler guessed (the fancy word is 'conjectured') that you need to add at least four fourth powers to get a fourth power, at least five fifth powers to get a fifth power, and so on.

Unlike Fermat, he was wrong. In 1966 Leon Lander and Thomas Parkin discovered that

$$27^5 + 84^5 + 110^5 + 133^5 = 144^5$$

This remained the only known example of the failure of Euler's conjecture until 1988, when Noam Elkies discovered that

$$2{,}682{,}440^4 + 15{,}365{,}639^4 + 187{,}960^4 = 20{,}615{,}673^4$$

In fact, Elkies proved that there are infinitely many cases where three fourth powers add up to a fourth power – but most of them require very big numbers. Roger Frye used a computer to search by trial and error, and found the smallest example:

$$95{,}800^4 + 217{,}519^4 + 414{,}560^4 = 422{,}481^4$$

● ●

The Millionth Digit

Suppose we write out all whole numbers in turn, strung together like this:

1234567891011121314151617181920212223242526...

and so on.

What is the millionth digit?

Answer on page 324

• •

Piratical Pathways

Roger Redbeard, the fiercest pirate in the Kidnibbean Sea, has
forgotten a vital piece of information – the address of his bank in
the Banana Islands, where he keeps his loot safe from the
attentions of the tax authorities. He knows which street it is on,
but there are more than thirty banks on Taxhaven Street, all
nameless, all looking exactly alike.

All is not lost, however, because he has a map.

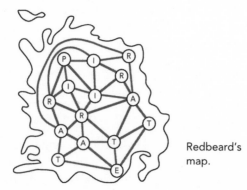

Redbeard's
map.

The address of his bank is cunningly concealed in this map: it
is the number of distinct ways to trace the word PIRATE, starting
at the circle marked P and spelling out the word letter by letter to
end at the circle marked E. The address is the number of different
ways that this can be achieved, always moving along the lines
linking the letters.

What is the address of Redbeard's Bank?

Answer on page 324

• •

Trains That Pass in the Siding

Two trains, the Atchison Flier (A) and the Topeka Bullet (B), are travelling in opposite directions towards each other along the same single-line track. Each consists of one locomotive, at the front, and nine coaches. Both locomotives and all coaches have the same length. The siding can accommodate no more than four coaches or locomotives in total at any one time, while leaving room for trains to pass along the main track.

Can the trains pass each other? If so, how?

Answer on page 325. [Hint: coaches can be decoupled.]

We're stuck – aren't we?

● ●

Please Make Yourself Clear

The mathematical logician Abraham Fraenkel, who was of German origin, once boarded a bus in Tel Aviv, Israel. The bus was scheduled to depart at 9.00 precisely, but by 9.05 it was still sitting in the bus station.

Aggrieved, Fraenkel waved a timetable at the driver.

'What are you – a German or a professor?' the driver enquired.

'Do you mean the inclusive or, or the exclusive or?' Fraenkel replied.*

Abraham Fraenkel.

● ●

* That is, do you allow *both* attributes, or only one?

Squares, Lists and Digital Sums

The list

$$81, \quad 100, \quad 121, \quad 144, \quad 169, \quad 196, \quad 225$$

consists of seven consecutive squares. It has a curious feature: the sum of the decimal digits of each of these numbers is itself a square. For example $1 + 6 + 9 = 16 = 4^2$.

Find another sequence of seven consecutive squares with the same property.

Answer on page 326

• •

Hilbert's Hit-List

In 1900, the German mathematician David Hilbert gave a famous lecture to the International Congress of Mathematicians in Paris, in which he listed 23 of the most important problems in mathematics. He didn't list Fermat's Last Theorem, but he mentioned it in the introduction. Here's a potted description of Hilbert's problems, and their current status.

1. *Continuum Hypothesis*
In Cantor's theory of infinite cardinal numbers (*Cabinet*, pages 157–61), is there a number strictly between the cardinalities of the integers and the real numbers?
   Solved by Paul Cohen in 1963 – the answer can go either way depending on which axioms you use for set theory.

2. *Logical Consistency of Arithmetic*
Prove that the standard axioms of arithmetic can never lead to a contradiction.
   Solved by Kurt Gödel in 1931, who proved that this can't be done with the usual axioms for set theory (*Cabinet*, page 205). On the other hand, Gerhard Gentzen proved in 1936 that it *can* be done using transfinite induction.

3. *Equality of Volumes of Tetrahedra*
If two tetrahedra have the same volume, can you always cut one into finitely many polyhedral pieces, and reassemble them to form the other?

Hilbert thought *not*. Solved in 1901 by Max Dehn – Hilbert was right.

4. *Straight Line as Shortest Distance Between Two Points*
Formulate axioms for geometry in terms of the above definition of 'straight line', and investigate what happens.

The problem is too broad to have a definitive solution, but much work has been done.

5. *Lie Groups Without Assuming Differentiability*
Technical issue in the theory of groups of transformations.

In one interpretation, solved by Andrew Gleason. However, if it is interpreted as the Hilbert–Smith conjecture,* it remains unsolved.

6. *Axioms for Physics*
Develop a rigorous system of axioms for mathematical areas of physics, such as probability and mechanics.

Andrei Kolmogorov axiomatised probability in 1933, but the question is a bit vague and is largely unsolved.

7. *Irrational and Transcendental Numbers*
Prove that certain numbers are irrational (not exact fractions) or transcendental (not solutions of polynomial equations with rational coefficients). In particular, show that, if a is algebraic and b is irrational, then a^b is transcendental – so, for example, $2^{\sqrt{2}}$ is transcendental.

Solved, affirmatively and independently, by Aleksandr Gelfond and Theodor Schneider in 1934.

---

* The group of p-adic integers has no faithful group action on a manifold. Hope that helps.

8. Riemann Hypothesis

Prove that all non-trivial zeros of Riemann's zeta function, in the theory of prime numbers, lie on the line 'real part $=\frac{1}{2}$'.

Unsolved. Possibly the biggest open problem in mathematics (see *Cabinet*, page 215).

9. Laws of Reciprocity in Number Fields

The classical law of quadratic reciprocity, conjectured by Euler and proved by Gauss in his *Disquisitiones Arithmeticae* of 1801, states that if p and q are odd primes then (see page 62 for notation) the equation $p \equiv x^2 \pmod q$ has a solution if and only if $q \equiv y^2 \pmod p$ has a solution, unless p and q are both of the form $4k - 1$, in which case one has a solution and the other does not. Generalise this to other powers than the square.

Partially solved.

10. Determine When a Diophantine Equation has Solutions

Find an algorithm which, when presented with a polynomial equation in many variables, determines whether any solutions in whole numbers exist.

In 1970, Yuri Matiyasevich, building on work by Julia Robinson, Martin Davis and Hilary Putnam, proved that there is no such algorithm.

11. Quadratic Forms with Algebraic Numbers as Coefficients

Technical issues, leading in particular to an understanding of the solution of many-variable quadratic Diophantine equations.

Partially solved.

12. Kronecker's Theorem on Abelian Fields

Technical issues generalising a theorem of Kronecker about complex roots of unity.

Still unsolved.

13. Solving Seventh-Degree Equations using Special Functions

Niels Henrik Abel and Évariste Galois proved that the general fifth-degree equation can't be solved using nth roots, but Charles Hermite showed that it can be solved using elliptic modular functions. Prove that the general seventh-degree equation can't be solved using functions of two variables.

A variant was disproved by Andrei Kolmogorov and Vladimir Arnold. Another plausible interpretation remains unsolved.

14. *Finiteness of Complete Systems of Functions*
Extend a theorem of Hilbert, about algebraic invariants for specific transformation groups, to all transformation groups.

Proved false by Masayoshi Nagata in 1959.

15. *Schubert's Enumerative Calculus*
Schubert found a non-rigorous method for counting various geometric configurations by making them as singular as possible (lots of lines overlapping, lots of points coinciding). Make this method rigorous.

Progress in special cases; no complete solution.

16. *Topology of Curves and Surfaces*
How many connected components can an algebraic curve of given degree, defined in the plane, have? How many distinct periodic cycles can an algebraic differential equation of given degree, defined in the plane, have?

Limited progress in special cases; no complete solution.

17. *Expressing Definite Forms by Squares*
If a rational function always takes non-negative values, must it be a sum of squares?

Solved by Emil Artin, D. W. Dubois and Albrecht Pfister. It is true over the real numbers, but false in some more general number systems.

18. *Tiling Space with Polyhedra*
General issues about filling space (Euclidean or not) with congruent polyhedra. Also mentions sphere-packing problems, notably the Kepler conjecture that the most efficient way to pack spheres in space is the face-centred-cubic lattice.

The Kepler problem has been solved, with a computer-aided

proof, by Thomas Hales (see *Cabinet*, page 231). The main question about polyhedra asked by Hilbert has also been solved.

19. *Analyticity of Solutions in Calculus of Variations*
The calculus of variations emerged from mechanics, and answers questions like: 'Find the shortest curve with the following properties.' If a problem in this area is defined by nice ('analytic') functions, must the solution be equally nice?

Proved by Ennio de Giorgi in 1957 and, with different methods, by John Nash.

20. *Boundary Value Problems*
Understand the solutions of the differential equations of physics, inside some region of space, when properties of the solution on the boundary of that region are prescribed. For example, mathematicians can find how a drum of given shape vibrates when its edge is fixed, but what if the edge is constrained in more complicated ways?

Essentially solved, by numerous mathematicians.

21. *Existence of Differential Equations with Given Monodromy*
A famous type of complex differential equation, called Fuchsian, can be understood in terms of its singular points and its monodromy group (which I won't even attempt to explain). Prove that any combination of these data can occur.

Answered yes or no, depending on interpretation.

22. *Uniformisation using Automorphic Functions*
Algebraic equations can be simplified by introducing suitable special functions. For instance, the equation $x^2 + y^2 = 1$ can be solved by setting $x = \cos \theta$ and $y = \sin \theta$ for a general angle θ. Poincaré proved that any two-variable algebraic equation can be 'uniformised' in this manner using functions of one variable. Technical question about extending these ideas to analytic equations.

Solved by Paul Koebe soon after 1900.

23. *Development of Calculus of Variations*

In Hilbert's day, the calculus of variations was in danger of becoming neglected, and he appealed for fresh ideas.

Much work has been done, but the question is too vague to be considered solved.

In 2000, the German historian Rüdiger Thiele discovered, in Hilbert's unpublished manuscripts, that he originally planned to include a 24th problem:

24. *Simplicity in Proof Theory*

Develop a rigorous theory of simplicity and complexity in mathematical proofs.

This is closely related to the concept of computational complexity, and the notorious (and unsolved) $P = NP$? problem (see *Cabinet*, page 199).

• •

Match Trick

Remove exactly *two* matches to leave two equilateral triangles.

Answer on page 327

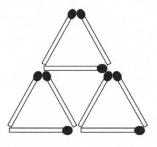

Take two matches away, and leave two triangles.

• •

Which Hospital Should Close?

Statisticians know that strange things happen when you combine data. One of them is Simpson's paradox, which I will illustrate with an example.

The Ministry of Health was collecting data on the success of surgical operations. Two hospitals – Saint Ambrose's Infirmary and Bumbledown General – were in the same area, and the ministry was going to close the less successful of the two.

- Saint Ambrose's Infirmary reported operating on 2,100 patients, of whom 63 (3%) died.
- Bumbledown General reported operating on 800 patients, of whom 16 (2%) died.

To the minister, the situation was perfectly obvious: Bumbledown General had a lower death rate, so he would close Saint Ambrose's Infirmary.

Naturally, the Chief Executive of Saint Ambrose's Infirmary protested. But he explained that there was a good reason for reconsidering, and asked the minister to break down the figures into two categories: male and female. The minister was reluctant to do so, on the grounds that it was obvious that Bumbledown General would still do better overall. However, it was easier to look at the new data than to argue, so he obtained the corresponding figures, classified by sex.

- Saint Ambrose's Infirmary operated on 600 females and 1,500 males. Of these, 6 females died (1%) and 57 males died (3.8%).
- Bumbledown General operated on 600 females and 200 males. Of these, 8 females died (1.33%) and 8 males died (4%).

Note that the numbers add up correctly, to give the original data.

Strangely, Bumbledown General had a worse death rate than Saint Ambrose's Infirmary in *both* categories. Yet, when the figures were combined, Saint Ambrose's Infirmary had a worse death rate than Bumbledown General.

In the end, the minister had to keep both hospitals open, unable to justify a decision either way if it were to be contested in court.

• •

How to Turn a Sphere Inside Out

In 1958, the distinguished American mathematician Stephen Smale, then a postgraduate student, solved an important problem in topology. But his theorem was so surprising that at first his thesis adviser Arnold Shapiro didn't believe it, pointing out that there was an obvious counterexample. That is, an example that proves the theorem false. One consequence of Smale's claimed result was that you can turn a sphere inside out using only continuous, indeed smooth, deformations. That is, you can't tear it, or cut holes in it, and you can't even make a sharp crease in it.

Intuitively, this seemed absurd. But intuition was wrong, and Smale was right.

Now, we all know that no matter how we twist and turn a balloon, the outside stays on the outside and the inside stays on the inside. Smale's work does not contradict this, because it permits one type of deformation that you can't do with a balloon. Namely, the surface is allowed to move through itself. However, it must do so in a smooth way, without creating a sharp crease. If creases are allowed, 'eversion' of the sphere, as it is called, is easy. Just push opposite hemispheres through each other, leaving a tube round the equator, and keep pushing so that the tube shrinks and disappears. However, this method creates an ever-sharper crease round the equator, and the technical definitions in Smale's theorem rule this out.

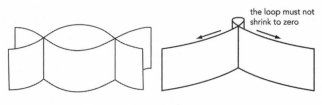

This is allowed ...   ... but this isn't.

So Smale was right, and the proof of his theorem could in principle be followed step by step to find an explicit method for everting a sphere. However, in practice this was too complicated, and for several years no specific method was known. The first method was devised by Shapiro and Anthony Phillips, and it was the first of what are now called *halfway models*.

Topologists have long known that some surfaces are 'one-sided'. The best known example is the Möbius band (*Cabinet*, page 111), and another is the Klein bottle (page 181). A sphere is two-sided: you can paint the inside surface red and the outside blue, say. But if you try to do that with a Möbius band or a Klein bottle, the red paint eventually runs into the blue paint: the apparent 'inside' and 'outside' surfaces in any small region connect together further round the band.

Now, there is another one-sided surface, the projective plane, which is closely related to a sphere. In fact, you can construct it mathematically by taking a sphere and pretending that diametrically opposite points are the same – in effect 'gluing' them together. The resulting surface can't be represented in three-dimensional space without passing through itself. But it can be 'immersed' in three-dimensional space, meaning that parts of it are permitted to pass smoothly through other parts.

Because the projective plane is a sphere with opposite points glued together, it can be pulled apart into a sphere by ungluing the pairs of points, which creates two separate layers, very close together. One of these is in effect the inside of the sphere, the other the outside. However, because the projective plane doesn't have an inside and outside, it can be pulled apart in two different ways. If we call the layers 'red' and 'blue', then the colours match up as the layers are pulled apart in the two different ways, but the red layer is inside for one way and outside for the other, while the blue layer is outside for one way and inside for the other.

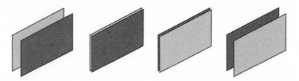

How red and blue layers interchange positions at the halfway stage.

The idea for a specific eversion, then, starts in the *middle* with an immersed projective plane. Pull it apart one way to create a sphere, with red on the outside and blue on the inside. Then deform that sphere, smoothly, until it looks like a normal round sphere, with only its red surface showing. This may not be easy, and it is not even obvious that it can be done, until you try. However, it works.

Now go back to the halfway stage, and pull the projective plane apart the *other* way, to create a sphere with blue on the outside and red on the inside. Then deform that sphere, smoothly, until it looks like a normal round sphere, with only its blue surface showing.

Fit these two deformations together by running the first one backwards. Now a sphere that is red on the outside and blue on the inside gets scrunged around, smoothly, until opposite pairs of points coincide at the midway projective plane. Pass the layers through each other, and pull them apart according to the second deformation. The result is a sphere that is blue on the outside and red on the inside.

Pull the projective plane apart two different ways ...

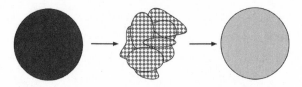

... then reverse the first deformation and combine the two.

Many different immersions of the projective plane are known. A famous one is Boy's surface. In 1901, the great German mathematician David Hilbert set his student Werner Boy a problem: to prove that the projective plane *can't* be immersed in three-dimensional space. Boy, like Smale, disagreed with his adviser. Like Smale, he was right. Boy had a surface named after him for his discovery.

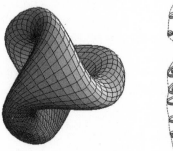

Boy's surface.

An advanced stage in the Shapiro–Phillips method.

A completely different method for turning a sphere inside out emerged from some general observations made by William Thurston, one of the world's greatest living geometers. Thurston devised a method in which the sphere is first corrugated, looking a bit like an exaggerated tangerine, with lots of segments poking out. This can be done by a smooth deformation. Then the north and south poles of the tangerine are pushed through each other, creating a series of handles round the equator. All the handles are simultaneously twisted through 180°. Then the north and south poles are pulled apart, creating another tangerine shape, but now the inside and outside of the original sphere have been swapped. It remains only to smooth away the corrugations.

Thurston's corrugation method.

All these methods for turning a sphere inside out are seriously complicated and difficult to follow, even with a lot of extra pictures and explanation. If you want to understand this topic fully, there is a wonderful video on:

www.youtube.com/watch?v=xaVJR60t4Zg

which you can download and watch to your heart's content. It was made by mathematicians at the Geometry Center at the University of Minnesota (unfortunately now closed), and it explains exactly how various sphere eversion methods work, with superb computer graphics. More information can also be found at:

www.geom.uiuc.edu/docs/outreach/oi/

Interestingly, you can't turn a *circle* inside out without creating creases – part of the intuition that made people think it was impossible for a sphere, too. This particular trick needs three dimensions to allow room to manoeuvre.

A Piece of String Walked into a Bar ...

A piece of string walked into a bar and ordered a beer.

'Sorry', said the barman. 'We don't serve strings.'

The string stomped out, muttering darkly about his funicular* rights. A little way up the street, he passed a stranger.

'You look like you could do with a beer,' said the stranger. 'It sure is hot.'

'I tried that, but the barman refused to serve me because I'm a string.'

'I can fix that,' said the stranger. He tied the string in a granny knot and frayed his ends. 'Try again.' So the string went back to the bar and asked again for a beer.

'Aren't you the piece of string that I just sent packing?' the barman asked suspiciously. 'You look just like him.'

'No,' the string replied. 'I'm a frayed knot.'

Slicing the Cake

If you cut a circular cake with 1, 2, 3 or 4 straight slices, then the largest number of pieces you can get is 2, 4, 7 and 11, respectively. (You're not allowed to move the pieces between cuts.)

What is the largest number of pieces you can create with five cuts?

Answer on page 328

The largest number of pieces with up to four cuts.

---

*   Look up the Latin *funiculus*.

The Origin of the Symbol for Pi

In 1647, the English mathematician William Oughtred wrote δ/π for the ratio of the diameter of a circle to its circumference. Here δ (Greek 'delta') is the initial letter of 'diameter', and π (Greek 'pi', of course) is the initial letter of 'perimeter' and 'periphery'. Isaac Barrow, another English mathematician, used the same symbols in 1664. The Scottish mathematician David Gregory (nephew of the famous James Gregory) similarly wrote π/ρ for the ratio of the circumference of a circle to its radius (ρ is the Greek 'rho', the initial letter of 'radius'). But to all these mathematicians, the symbols referred to different lengths, depending on the size of the circle.

In 1706, the Welsh mathematician William Jones used π to denote the ratio of the circumference of a circle to its diameter, in a work that gave the result of John Machin's calculation of π to 100 decimal places.

In the early 1730s, Euler used the symbols p and c, and history might have been different, but, in 1736, he changed his mind and started to use the symbol π in its modern sense. It came into general use after 1748, when he published his *Introduction to the Analysis of the Infinite*.

• •

Hall of Mirrors

If someone lights a match in a hall of mirrors, can it be seen (reflected as many times as necessary) from any other location?

Let me make the question precise. We restrict attention to two dimensions of space – the plane. Recall that when a light ray hits a flat mirror, it bounces off again at the same angle. Suppose you have a room – a polygonal region – in the plane, whose boundary consists of flat mirrors. A point source of light is placed somewhere in the interior of the room. Can this source always be seen, perhaps after multiple reflections, from any other interior point? Light that hits any corner of the polygon is absorbed and stops.

Victor Klee published this question in 1969, but it goes back to Ernst Straus in the 1950s, if not earlier. In 1958, Lionel and Roger Penrose found a room with a curved edge for which the answer is 'no', but the question for polygons remained open until George Tokarsky solved it in 1995. Again, the answer is 'no'. He found many rooms with that property: the picture shows one of them. It has 26 sides and every corner lies on a square grid.

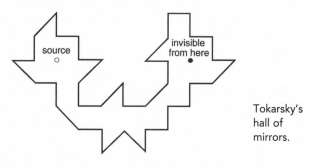

Tokarsky's hall of mirrors.

Greek and Trojan Asteroids

Two unusual clumps of asteroids occupy much the same orbit as Jupiter. Unlike the 'clumps' in the asteroid belt (page 120), these clumps really are clumps – the asteroids stay together in a cluster. Though they are still separated by huge distances: space is *big*. One clump, the Greeks, is spread out around a position 60° ahead of Jupiter; the other clump, the Trojans, lags 60° behind it. The individual asteroids are (mostly) named after characters in Homer's *Iliad*, a story of the siege of Troy by the Greeks, belonging to the appropriate sides.

The discovery of the Trojans in the 1900s confirmed a prediction that the Italian-born mathematician Joseph Louis Lagrange made in 1772. He worked out the combined effects of gravity and centrifugal force in a miniature solar system containing a sun and one planet, in a circular orbit. The same goes for any two-body gravitational system with a circular orbit, such as the Earth and the Moon – to a good approximation, at least. His calculations showed that there are exactly five points, relative to these two bodies, at which gravity and centrifugal force cancel out exactly, so that a small mass located at such a point will stay in equilibrium. These are the *Lagrangian points* L_1–L_5.

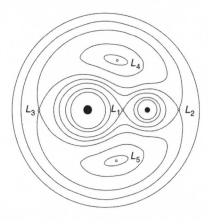

Lagrangian points, and associated energy contours.

- L_1 lies between the Sun and the planet.
- L_2 lies on the far side of the planet, along a line joining the Sun and the planet.
- L_3 lies on the far side of the Sun, along a line joining the Sun and the planet.
- L_4 lies in the planet's orbit, 60° ahead of it.
- L_5 lies in the planet's orbit, 60° behind it.

More precisely, around 1750, Leonhard Euler proved that the points L_1, L_2 and L_3 exist, and Lagrange discovered the other two. Lagrange did this calculation as part of an attack on a more general question, the motion of three bodies under gravity. Isaac Newton had shown that, for two bodies, the orbits are ellipses, and it was natural to ask what happens with three bodies. This turned out to be a very difficult problem, and we now know why: the typical motion is chaotic (*Cabinet*, page 117).

The L_4 and L_5 points are stable, provided the mass of the Sun is at least

$$\frac{25 + 3\sqrt{69}}{2} \approx 24.96$$

times that of the planet. That is, a mass located at such a point will remain nearby even if it is disturbed a little. The other three points are unstable. No natural occurrences of bodies orbiting at these points were known until astronomers noticed that unusually many asteroids are located near the Sun–Jupiter L_4 and L_5 points. They are spread out along Jupiter's orbit in the same 'banana' shape as the energy contours near those points. Since then, other instances have been found:

- The Sun–Earth L_4 and L_5 points contain interplanetary dust.
- The Earth–Moon L_4 and L_5 points may contain interplanetary dust in so-called Kordylewski clouds.
- The Sun–Neptune L_4 and L_5 points contain Kuiper belt objects, a class of smallish bodies now including Pluto, most of which orbit further out than Pluto.
- The Saturn–Tethys L_4 and L_5 points hold the small moons Telesto and Calypso.
- The Saturn–Dione L_4 and L_5 points hold the small moons Helene and Polydeuces.

Although the other three Lagrangian points are unstable, they are surrounded by stable orbits, called halo orbits, so a space probe or other artefact can be maintained near those points with very little expenditure of fuel. The James Webb Space Telescope, successor to the Hubble Telescope, will be positioned at the Sun–Earth L_2 point when it is launched in or after 2013. This location keeps the Earth and Sun in the same direction, as seen from the telescope, so that a single fixed shield can stop radiation from those two bodies warming it up and disturbing the delicate instruments. The only Lagrangian point that has not yet featured in an actual or planned space mission is L_3. All five of them have been exploited in numerous science fiction stories.

A wealth of further information can be found at:
en.wikipedia.org/wiki/Lagrangian_point

Sliding Coins

A good pub puzzle. Start with six coins, numbered 1–6 and arranged as in the left-hand picture. Slide them one at a time, without disturbing the others, to rearrange them into the right-hand picture in the number order shown.

How can you achieve this by moving as few coins as possible?

Answer on page 328

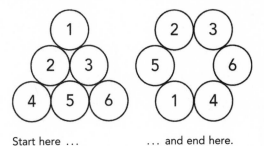

Start here ...   ... and end here.

Beat That!

... and *then* what?

Chapter 94 of Snorri Sturluson's *Heimskringla: History of the Kings of Norway* – which I'm sure you're familiar with – tells of a game of chance between King Olaf I of Norway* and the King of Sweden,† to decide which country owned the island of Hísing.

According to Thorstein the Learned, the two kings agreed to throw a pair of dice, and whoever got the highest score also got the island.

The King of Sweden, who had won the right to go first by drawing lots, threw the dice, and scored a double six. 'There is no use in you throwing,' he said. 'I cannot lose.'

'There remain two sixes on the dice, my Lord,' replied Olaf, as he shook the dice in his hand, 'and it is a trifling matter for God to make the dice land that way.' Then he rolled the dice ...

What do you think happened next?

Answer on page 328

• •

Euclid's Puzzle

Legend has it that the great geometer Euclid composed a puzzle which went as follows.

A mule and a donkey were stumbling along the road, each carrying several identical heavy sacks. The donkey started complaining, making a horrible groaning noise, and eventually the mule got fed up.

'What are *you* complaining for? If you gave me one sack, I'd have twice as many as you! And if I gave you one sack, we'd be carrying the same load.'

How many sacks were the donkey and the mule carrying?

Answer on page 329

• •

* That is, Olaf Tryggvason, the son of Tryggve Olafsson, who was king from 995 to 1000. Before the game of dice, Olaf had proposed marriage to Sigrid the Haughty, the Queen of Sweden, in an attempt to unite Scandinavia. She wasn't keen.
† Who seems to have been Olof the Treasurer, from the dates. As it happens, he was the son of Eric the Victorious and Sigrid the Haughty. It's a small world.

The Infinite Monkey Theorem

It is said that if a monkey sat at a typewriter and kept hitting keys at random, then eventually it would type the complete works of Shakespeare. This statement dramatises two things about random sequences: anything can turn up, and, therefore, the result *need not appear random*. The infinite monkey theorem goes further, and states that, if the monkey keeps typing for ever, then the probability that it will eventually type any given text is 1.

To test this proposition, all you need is two dice, of different colours or otherwise distinguishable, and a table of symbols. The one at bottom right is a space.

	first dice					
	1	2	3	4	5	6
1	A	B	C	D	E	F
2	G	H	I	J	K	L
3	M	N	O	P	Q	R
4	S	T	U	V	W	X
5	Y	Z	.	,	:	;
6	'	"	-	?	!	

second dice

Simulated monkey.

Throw the two dice, choose the corresponding symbol, and write it down. For instance, if you throw 4 /1, then you get the letter D. Keep going, and see how long it takes to get a sensible word with, say, three or more letters. Your experience should be confirmed by two calculations:

- On average, how many throws would it take to get DEAR SIR, including the space between the words?
- On average, how many throws would it take to get the complete works of Shakespeare? You may assume that his works contain 5,000,000 characters, all included in the table. It's not true, but assume it anyway.

Answers on page 329

In 2003, lecturers and students from the University of Plymouth MediaLab tried the experiment with real monkeys –

six Celebes crested macaques – and a computer keyboard. The experimental subjects produced five pages of typing, mainly looking like this:

SSSSSSSSSSSSSSSSSSSSSSSSSSSSSSSSSSSSSSSS

and then trashed the keyboard comprehensively.

The mathematical statement goes back to Émile Borel, in a 1913 paper 'Statistical mechanics and irreversibility', and his 1914 book *Le Hasard* (*Chance*). The Argentine writer Jorge Luis Borges traced the underlying idea back to Aristotle's *Metaphysics*. The Roman orator Cicero, unimpressed by Aristotle's views, compared the statement to believing that 'if a great quantity of the one-and-twenty letters, composed either of gold or any other matter, were thrown upon the ground, they would fall into such order as legibly to form the *Annals of Ennius*. I doubt whether fortune could make a single verse of them.'

Well, no ... unless you used a *really* great quantity.

● ●

Monkeys Against Evolution

The monkey on the typewriter has been used to attack the theory of evolution.* Random mutations in DNA are like the monkey. And while it is true that *eventually* the monkey can type anything, it is also true that it won't type anything remotely interesting during the lifetime of the universe. Now, a key protein like haemoglobin, which carries oxygen in our blood, is specified by more than 1,700 DNA 'letters' A, C, T, G. The chance of this molecule arising by random mutations is so tiny that it might as well be zero. Therefore haemoglobin cannot have evolved, Darwin was wrong, God must have created it, QED.

This criticism turns out to be facile, and rests on several

---

* According to which a monkey *did* write the collected works of Shakespeare, though not – to begin with – on a typewriter. It performed the feat indirectly, by producing descendants that evolved into ... Shakespeare. This is a far more efficient approach.

misconceptions. One is that the haemoglobin molecule is a 'target' at which evolution must aim. However, haemoglobin is not the only molecule that could carry oxygen and deliver it where required. Haemoglobin does that job because it has two similar but distinct forms. In one of them, oxygen atoms bind to the four iron atoms in the molecule; in the other, they don't. The molecule 'flexes' slightly from one form to the other. Most of the haemoglobin molecule plays no essential role in this process, although it does provide a suitably flexible scaffolding for the bits that matter. So a huge variety of other molecules could in principle do the same job. Nature evolved one, and that was all it needed. Well, actually it evolved several variants, which if anything helps to support the point I'm making.

That point alone doesn't cut the odds down enough, though. The second point is that biological molecules don't evolve from scratch every time: evolution keeps a living library of molecules, and modifies them or fits them together to build new ones. Indeed, haemoglobin is made from two copies of each of two smaller molecules, alpha and beta units. Moreover, this modular structure helps the combined molecule to flex appropriately.

A more appropriate analogy, then, equips the monkey with a wordprocessor, not a typewriter, and the wordprocessor has 'macro' keys, which can be assigned to reproduce a series of keystrokes. If the monkey creates a macro every time it types a sensible word – analogous to evolution keeping anything that works – then pretty soon the monkey's computer will build up a dictionary, and can type sequences of words with ease by concentrating on the macro keys. Repeating the process produces sequences of meaningful sentences, and so on. It might not generate Shakespeare, but in a few years, let alone billions, a monkey with macros could put together an article that you could read on the train.

That said, evolving something to play the role of haemoglobin takes a long time, even when gigantic numbers of molecules play the game in parallel – as they do today, and presumably did in the distant past. It took about 3 billion years for haemoglobin to evolve. However, for much of that time, it wouldn't have had any useful function – complex creatures able to survive in a toxic oxygen atmosphere did not arrive until 1.5 billion years or so had passed, and blood cells arose a lot later than that – and it turned up fairly rapidly, by geological standards, once the scene was set for it to do something useful. But it did so through a sequence of processes that combined small molecules into bigger ones, then those into bigger ones still. It didn't just faff around at random hoping to hit the haemoglobin jackpot by choosing the right 1,700 DNA letters.

• •

Universal Letter of Reference

Dear Search Committee Chair,

I am writing this letter for Mr XXXXX, who has applied for a position in your department.

I should start by saying that I cannot recommend him too highly.

In fact, there is no other student with whom I can adequately compare him, and I am sure that the amount of mathematics he knows will surprise you.

His dissertation is the sort of work you don't expect to see these days. It definitely demonstrates his complete capabilities.

In closing, let me say that you will be fortunate if you can get him to work for you.

Sincerely,

A. D. Visor (Prof.)

From *Focus* Newsletter, Mathematical Association of America.

• •

Snakes and Adders

This is a playable game for two or more players with topological and combinatorial features. It is a slight modification of a game that Larry Black invented in 1960, called the *Black Path Game*.

Start by drawing a grid on paper; 8×8 is about right. Draw a cross at top left. Remove the diagonally opposite corner square – I'll explain why in a moment.

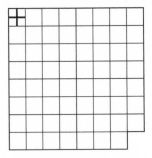

Starting position
for the game.

The first player draws one of the following symbols in the square next to the + sign, horizontally or vertically:

Symbols to be drawn.

Players then take turns to draw one of the three symbols – whichever they prefer – in the unique square that extends the wiggling 'snake' started by the first player. The snake can overlap itself at a + symbol.

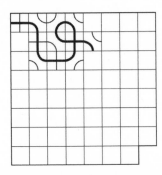

State of the game
after a few moves.
The snake is the heavy line.

Whoever first makes the snake run into the edge of the board, including the indentation at bottom right, loses. The topology of the snake implies that it can't stop at an interior point of the big square, and it can't run into a closed loop. So it must eventually terminate at the edge.

This game is fun to play, and you may wonder what that excised corner square is all about. If you don't cut out the corner square, but use the full 8×8 board, there is a simple winning strategy for one of the players.

Who should win, and how?

Answer on page 330

- -

Powerful Crossnumber

Fill in the
eight powers.

Here's a crossnumber with a difference – I'm not going to give you the clues. But I will tell you that each of the answers (2, 5, 6, 7 across; 1, 2, 3, 4 down) is a power of a whole number, and the answers comprise two squares, one cube, one fifth power, one sixth power, one seventh power, one ninth power and one twelfth power.

Now, a sixth power is also a cube and a square, because $x^6 = (x^2)^3 = (x^3)^2$. To avoid ambiguity, when I say that a solution is some specific power, I mean that it is *not* also some higher power. And there should be no leading zeros – so 0008, for instance, does not count as the cube of 2.

Answer on page 331

- -

Magic Handkerchiefs

A professional magician like the Great Whodunni is never
without a handkerchief or ten, and can produce them indefinitely
from a top hat, a sealed and empty box, or a volunteer's pockets.
Sometimes the odd pigeon turns up too, but to emulate this
particular trick (which Whodunni learned from the American
magician Edwin Tabor) all you need is two handkerchiefs –
preferably of different colours. Roll up each along its diagonal to
make a thick roll of cloth about a foot (30 cm) long.

Now follow the instructions and pictures.

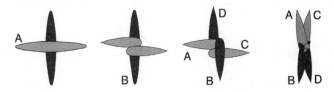

Handkerchief trick.

1   Cross the handkerchiefs with the dark one underneath.
2   Reach under the dark handkerchief, grab end A of the light
handkerchief, pull it behind the dark handkerchief, and wrap it
over the front of the dark handkerchief.
3   Reach under the light handkerchief, grab end B of the dark
handkerchief, pull it behind the light handkerchief, and wrap it
over the front of the light handkerchief.
4   Bring ends B and D together by swinging them underneath
the rest of the handkerchief. Bring ends A and C together by
swinging them over the top of the rest of the handkerchief.

Now the two handkerchiefs are all tangled together. Hold ends A
and C together in one hand, and B and D together in the other
hand. Now pull your hands quickly apart.

What happens?

Answer on page 331

A Bluffer's Guide to Symmetry

The word 'symmetry' is often bandied about, but in mathematics it has a precise – and very important – meaning. In everyday language, we say that an object is symmetrical if it has an elegant shape, or is well proportioned, or (getting technical) the left and right sides of the object look the same. The human figure, for instance, looks much the same when reflected in a mirror.

The mathematical usage of the word 'symmetry' is significantly different and much broader: mathematicians talk of '*a* symmetry' of an object, or '*many* symmetries'. To mathematicians, a symmetry is not a number, or a shape, but a *transformation*. It is a way to move an object, so that when you've finished, the object appears not to have changed.

The cat (far left) looks different if you rotate it ...

... or reflect it ...

... so it has no symmetries.
No, that's a lie: it has *one* symmetry:
leave it alone. This is the *trivial
symmetry*, and all shapes have it.

A cat with two tails looks the same when you reflect it, so it has an axis of reflectional symmetry (grey line).

The cat's body has two axes of reflectional symmetry, and it also looks the same when you rotate it through 180°.

Four cats sitting in a square are symmetric under rotations of 0° (trivial), 90°, 180° and 270°. This is 4-fold rotational symmetry.

The same goes when you throw away the cats …

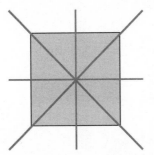

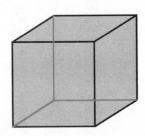

… but now there are four new axes of reflectional symmetry. So a square has *eight* different symmetries.

A cube has 48 symmetries …

... and a dodecahedron has 120.   A circle has infinitely many rotational symmetries (any angle) and infinitely many reflectional symmetries (any diameter as axis).

If this line of cats went on infinitely far, it would have *translational* symmetries: slide the cats an integer number of spaces right or left.

A cat crystal has translational symmetries in two different directions.

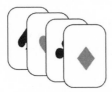

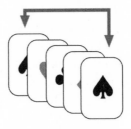

Symmetries need not be motions. Shuffling a pack of cards is a transformation ...

... and if some cards are identical, some shuffles just swap identical cards – these are permutational symmetries of the pack.

Symmetries have come to dominate huge areas of mathematics. They are very general – it's not only shapes that have symmetries. So do number systems, equations, and processes of all kinds. The symmetries of a mathematical 'thing' tell us a lot about it. For instance, Galois proved that you can't solve the general equation of the fifth degree by an algebraic formula, and the main point of his proof is that the general equation of the fifth degree has *the wrong kind of symmetries*.

Symmetries are vital in physics, too. They classify the atomic lattices of crystals – there are 230 different symmetry types, or 219 if you consider mirror images to be the same. The 'laws of nature' turn out to be highly symmetric, mainly because the same laws operate at all points of space and all instants of time. The symmetries of the laws tell us a lot about the solutions. Quantum physics and relativity are both based on symmetry principles.

Front–back symmetry of a pacing giraffe. The front and back legs on each side hit the ground together.

Symmetries are even turning up in biology. Many important biological molecules are symmetric, and the symmetries affect how they work. But you can find symmetries in the shapes of animals, in their markings, and even in how they move. For example, when a giraffe paces, it moves both left legs together, then both right legs together. So the front legs do the same as the back legs, like two people walking one behind the other, in step with each other. The symmetry here is a permutation: *swap front and back.*

Perform this only in the abstract, please, or the giraffe will get upset.

• •

Digital Century Revisited

Innumeratus wrote the nine non-zero digits down in order, with gaps, like this:

$$1\ 2\ 3\ 4\ 5\ 6\ 7\ 8\ 9$$

'I want you to ... ' he began.

'... make 100 by inserting standard arithmetical symbols,' said Mathophila. 'That's easy, it was in *Professor Stewart's Cabinet of Mathematical Curiosities*, which you gave me for Christmas, but it goes back a lot further than that.' And she wrote:

$$123 - 45 - 67 + 89 = 100$$

'No, that's cheating,' said Innumeratus. 'I left *gaps*! You can't consider 1 2 3 to be one hundred and twenty-three, and ... '

'Oh. No concatenation of symbols allowed, then.'

'Yeah. No caterwaulification ... whatever.'

She thought for a moment, and wrote down

$$(1 + 2 - 3 - 4) \times (5 - 6 - 7 - 8 - 9)$$

'Sorry, no brackets,' said Innumeratus.

Mathophila shrugged, and wrote

$$1 + 2 \times 3 + 4 \times 5 - 6 + 7 + 8 \times 9$$

'You don't mind me using the rule that multiplication precedes addition, so I don't need to put brackets round individual multiplications, do you?'

'No, that's OK. But ... uh ... look, sorry, but no subtraction symbols either.'

There was a silence. 'I'm not sure that's possible,' said Mathophila.

'Wanna bet?' asked Innumeratus smugly.

What should Mathophila do?

Answer on page 331

Answer on page 331

● ●

An Infinity of Primes

Euclid proved that there is no largest prime. Here's a quick way to see this: if p is prime then $p! + 1$ is not divisible by any of the numbers 2, 3, ..., p, since any such division leaves remainder 1. So all its prime factors are bigger than p. Here, $p! = p \times (p - 1) \times (p - 2) \times \cdots \times 3 \times 2 \times 1$.

Euclid's proof was slightly different. He stated it geometrically, and in modern terms he used a typical example to show that if you have any finite list of primes, then you can get a bigger one by multiplying them all together, adding 1, and then taking any prime factor of the result.

This suggests an interesting sequence of primes, all guaranteed to be different:

$$p_1 = 2$$
$$p_2 = 3$$
$$\vdots$$

$p_{n+1} =$ the *smallest* prime factor of $p_1 \times p_2 \times \cdots \times p_n + 1$

For example,

$p_3 =$ the smallest prime factor of $2 \times 3 + 1 = 7$, namely 7
$p_4 =$ the smallest prime factor of $2 \times 3 \times 7 + 1 = 43$, namely 43

$p_3 =$ the smallest prime factor of $2 \times 3 \times 7 \times 43 + 1 = 1807$,
    namely 13

(because $1807 = 13 \times 139$), and so on.

The first few terms are

2, 3, 7, 43, 13, 53, 5, 6221671, 38709183810571, 139, 2801, 11, 17, 5471, 52662739, 23003, 30693651606209, 37, 1741, 1313797957

and the sequence is highly irregular. Occasionally the product $p_1 \times p_2 \times \cdots \times p_n + 1$ is prime, and the size goes up enormously, but when it's not prime, the smallest factor is often very small indeed. This behaviour is pretty much what you might expect, wild though it may be.

Despite (or perhaps because of) this tendency to swing madly between huge numbers and tiny ones, the first 13 terms include the first seven primes: 2, 3, 5, 7, 11, 13, 17. Which raises an interesting – and probably difficult – question: does *every* prime occur somewhere in this sequence?

I have no idea how to answer that, though if I had to guess I'd say it's true.

. . . .

A Century in Fractions

The famous English puzzlist Henry Ernest Dudeney remarked that the fraction

$$91\frac{5{,}742}{638}$$

is equal to 100, and uses every digit 1–9 exactly once. He found ten other ways to achieve this, one of which has only one digit before the fractional part. What was this solution?

Answer on page 332

. . . .

Ah, That Explains It . . .

- Knowledge is power
- Time is money

But, by definition,

- Power = work/time

So,

- Time = work/power

which implies that

- Money = work/knowledge

Therefore:

- For a fixed amount of work, the more you know, the less money you get.

● ●

Life, Recursion and Everything

Readers of Douglas Adams's *The Hitch Hiker's Guide to the Galaxy* will recall the prominent role of the number 42 – the answer to the Great Question of Life, the Universe and Everything. The question turned out to be 'what is six times nine?', which was vaguely disappointing. Anyway, Adams chose 42 because a quick poll of his friends suggested that this was the most boring number they could think of.

It's true that interesting properties of 42 don't exactly trip off the tongue, but we know (*Cabinet*, page 105) that all numbers are interesting. However, the proof is non-constructive. So I was pleased to find out about a natural occurrence of 42 as an interesting number. It arises in a sequence of numbers introduced by F. Göbel. Suppose we define

$$x_0 = 1$$

$$x_1 = \frac{1 + x_0^2}{1}$$

$$x_2 = \frac{1 + x_0^2 + x_1^2}{2}$$

$$\vdots$$

$$x_n = \frac{1 + x_0^2 + x_1^2 + \cdots + x_{n-1}^2}{n}$$

There is no obvious reason why the x_n should be whole numbers, but the first few terms of the sequence are

  1, 2, 3, 5, 10, 28, 154, 3520, 1551880, 267593772160

so you do begin to wonder whether, by some miracle, all the terms are integers.

The truth is, if anything, even more miraculous. Hendrik Lenstra put the equation on a computer, and discovered that the first term that is not an integer is x_{43}. So 42 is the largest integer for which all terms of the sequence, up to and including that one, are integers.

Other sequences of this kind also seem to behave that way – a lot of integers to begin with, but at some point the pattern fails. With the same rule but using sums of cubes, the first term that is not an integer is x_{89}. With fourth powers the first non-integer is x_{97}, with fifth powers it is x_{214}, with sixth powers it is the relatively feeble x_{19}, but with seventh powers we get the astonishing x_{239}. So here is a sequence with a nice pattern, such that the first 238 terms* are integers, but the 239th is not.

As far as I know, no one really understands why these sequences behave like they do.

• •

* I'm not counting x_0 here, although that's an integer too. However, it's an arbitrary starting-point, which is a reason – not a terribly good one, but a reason nonetheless – for omitting it from the count. I mention this only because dozens of readers will write to me about it if I don't. Anyway, if I included x_0, then 42 would become 43, and the gratuitous link to *Hitch Hiker* wouldn't work.

False, Not Stated, Not Proved

James Joseph Sylvester was a 19th-century mathematician who specialised in algebra and geometry. He worked a lot of the time with Arthur Cayley, whose day job was in the law. Cayley had a superb memory and knew almost everything that was going on in mathematics. Sylvester was the exact opposite.

On one occasion the American mathematician William Pitt Durfee sent some of his work to Sylvester, only to be informed that the first theorem in it was false, and had never even been stated, let alone proved. Durfee produced a paper whose main objective was to prove the theorem concerned, which it did successfully.

The paper had been written by Sylvester.

James Joseph Sylvester.

● ●

Proof That 2 + 2 = 4

By definition,

$$2 = 1 + 1$$
$$3 = 2 + 1$$
$$4 = 3 + 1$$

Therefore,

$$2 + 2 = (1 + 1) + (1 + 1)$$
$$= ((1 + 1) + 1) + 1 \qquad (*)$$
$$= (2 + 1) + 1$$
$$= 3 + 1$$
$$= 4$$

where (*) is justified by the associative law

$$(a + b) + c = a + (b + c)$$

with $a = (1 + 1)$, $b = 1$, $c = 1$.

See Note on page 332

• •

Slicing the Doughnut

If you cut this doughnut with three straight slices, what is the largest number of pieces you can create? (You're not allowed to move the pieces between cuts.)

Answer on page 332

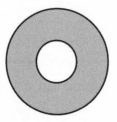

How many pieces
with three cuts?

• •

The Kissing Number

If you try to surround a circular coin by coins of the same type, so that all the other coins touch the first one, you quickly discover that exactly six coins fit neatly round the first.

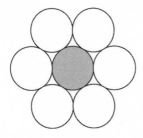

In 2 dimensions the kissing number is 6.

This isn't exactly news to most of us, but it leads to a concept that turns out to be important in the theory of digital codes, as well as having mathematical interest in its own right. A coin is a circle, which is a 2-dimensional shape, so we've just seen that the *kissing number* in 2-dimensional space is 6. In n-dimensional space, the kissing number is similarly defined to be the largest number of non-overlapping unit $(n-1)$-spheres that can touch ('kiss') a unit $(n-1)$-sphere. Here an $(n-1)$-sphere is the natural analogue of a circle (1-sphere) or sphere (2-sphere). The number drops from n to $n-1$ because although a sphere, say, lives in 3-dimensional space, its surface has only 2 dimensions. And a circle is a curve (hence 1D) in a 2D space, the plane. The unit $(n-1)$-sphere, in fact, comprises all points in n-dimensional space that are distance 1 from some fixed point, the centre of the $(n-1)$-sphere.

The exact value of the kissing number is known for very few dimensions: 1, 2, 3, 4, 8 and 24, in fact. In 1D space, which is a line, a 0-sphere is a pair of points spaced 2 units apart (the *diameter* of a unit n-sphere is 2). So the kissing number in 1D is 2: one on the left, one on the right. We've just seen that in 2D space the kissing number is 6. What about higher dimensions?

In 3D space, it is easy to get 12 spheres to kiss a single sphere: you can do it with ping-pong balls and glue dots. But the arrangement is 'sloppy', with room to move the spheres around, and with quite a bit of space left between them. Can you fit in a 13th sphere? In 1694, David Gregory, a Scottish mathematician, thought it could be done; no lesser a luminary than Isaac Newton disagreed. The issue was sufficiently delicate that it was not resolved until 1874; it then turned out that Newton was right. So the kissing number in 3D space is 12.

In 3 dimensions the kissing number is 12.

A similar story holds in 4D space, where it's relatively easy to find an arrangement of 24 kissing 3-spheres, but there's enough room left so that maybe a 25th might fit in. This gap was eventually sorted out by Oleg Musin in 2003: the answer is 24.

In most other dimensions, mathematicians know that some particular number of kissing spheres is possible, because they can find such an arrangement, and that some generally much larger number is impossible, for various indirect reasons. These numbers are called the lower bound and upper bound for the kissing number, and it must lie between them.

In just two cases beyond 4D, the known lower and upper bounds coincide, and their common value is therefore the kissing number. These dimensions are 8 and 24, where the kissing numbers are, respectively, 240 and 196,650. In these dimensions there exist two highly symmetric lattices, higher-dimensional analogues of grids of squares or more generally grids of parallelograms. These special lattices are known as E_8 (or the Coxeter–Todd lattice) and the Leech lattice, and spheres can be placed at suitable lattice points. By an almost miraculous coincidence, the provable upper bounds for the kissing number in these dimensions are the same as the lower bounds provided by these special lattices.

The current state of play can be summed up in a table, where

I've used boldface for those dimensions where an exact answer is known:

Dimension	Lower bound	Upper bound	Dimension	Lower bound	Upper bound
1	**2**	**2**	13	1,130	2,233
2	**6**	**6**	14	1,582	3,492
3	**12**	**12**	15	2,564	5,431
4	**24**	**24**	16	4,320	8,313
5	40	45	17	5,346	12,215
6	72	78	18	7,398	17,877
7	126	135	19	10,688	25,901
8	**240**	**240**	20	17,400	37,974
9	306	366	21	27,720	56,852
10	500	567	22	49,896	86,537
11	582	915	23	93,150	128,096
12	840	1,416	**24**	**196,560**	**196,560**

The best known lower bounds, for all dimensions up to 40 and a few larger ones, can be found at:
www.research.att.com/~njas/lattices/kiss.html

The kissing number for regular arrangements, in which the centres of the spheres all lie on a lattice, is known exactly for 1–9 dimensions, as well as 24. In 1, 2, 3, 4, 8 and 24 dimensions, it is the value shown in the table. For 5, 6, 7 and 9 dimensions it is, respectively, 40, 72, 126 and 272. (The table entry 306 in 9 dimensions does not refer to a regular arrangement.)

● ●

Tippe Top Twister

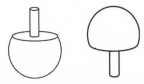

The two positions of a tippe top.

The toy known as a tippe (or alternatively 'tippy') top is made by slicing a bit off a sphere and adding a cylindrical 'stalk'. When you spin it – fast enough – it turns upside down. Most of us have played with a tippe top at some point, but here's a question we possibly haven't thought about. Suppose that, when you first spin the top, while it is still the usual way up, you spin it clockwise, looking down from above. This is the natural direction for right-handers.

When it turns over, in which direction does it spin?

Answer on page 333

When Is a Knot Not Knotted?

Topologists study things like knots, and they try to work out whether two knots are 'topologically equivalent', that is, can be deformed into each other. Or not. To do that, they invent cunning 'invariants', which are equal for equivalent knots, but may or may not be equal for two knots that are not equivalent. So knots with different invariants are definitely topologically different, but knots with the same invariants may or may not be topologically different.

It's a knotty issue. Most of the useful invariants aren't perfect: they're a bit like using 'odd/even' to distinguish people's ages. If Eva's age is even and Ollie's age is odd, then we know their ages must be different, even if we don't know what their ages *are*. But if Evangeline's age is even and Everett's age is even, then their ages might be the same (for instance, 24 and 24) or maybe not (24 and 52). So in this case we can't tell.

Sometimes topologists get lucky, and the invariant is good enough to tell them when a knot is not actually knotted, even if it can't reliably distinguish all different knots. A case in point is the so-called 'knot group', one of the first knot invariants discovered. I mention all this not because of the topology, which is highly technical, but because in 1972, in the mathematical fanzine *Manifold*, it gave rise to a poem that summed up what was good and bad about the knot group. It bore the title *Knode*:

A knot and
Another
knot may
not be the
same knot, though
the knot group of
the knot and the
other knot's
knot group
differ not; BUT
if the knot group
of a knot
is the knot group
of the not
knotted
knot, then
the knot is
not
knotted.

* *

The Origin of the Factorial Symbol

The early symbol for 'factorial n', which is

$$n \times (n-1) \times (n-2) \times \cdots \times 3 \times 2 \times 1$$

was

$$n\,|$$

but this was tricky to print. So in 1808, the French mathematician Christian Kramp decided to change it to

$$n!$$

which was easy to typeset. The old-fashioned version quickly went out of vogue, one of a number of examples where the practicalities of printing have affected mathematical symbolism.

* *

Juniper Green

'Let's play a number game,' said Mathophila.

Innumeratus, ever a sucker, took the bait. 'What kind of game?'

Mathophila placed cards numbered 1–100 face up on the table.* 'I'll show you the rules.' She wrote down:

- Players take turns to choose one card. The chosen card is removed and cannot be used again.
- Apart from the opening move, the chosen number must either be an exact divisor of the previous one, or an exact multiple of it.
- The first player who cannot obey the rules loses.

'OK,' said Innumeratus. 'You go first.'

'Well, actually—' Mathophila began, but then stopped. 'Oh, very well.' She picked up card 97 and discarded it.

Innumeratus, after some counting on fingers, said 'That's prime, isn't it?' When she nodded, he added 'So I've got to choose card 1.'

'Yes. The only other divisor is 97, and that's gone. The smallest multiple is 194, and that's too big.'

So Innumeratus picked up card 1, and discarded it.

Mathophila grinned, and picked up 89. 'You lose.'

'That's prime, too?' asked Innumeratus, who could sometimes be quite bright.

'Yes.'

'So I have to choose 1 again ... Oh. I can't, it's already gone.' He paused. 'That's a silly game. The first player always wins.'

'Yes, we call it the double-whammy tactic.'

Innumeratus thought for a moment. 'OK, let me start. Now I'll choose a prime.' And he picked up card 47.

Mathophila, disdaining the 1, chose 94 instead.

---

* To play the game you will need to make a set of cards – I don't know anywhere that sells them. It's worth the effort.

'Oops,' said Innumeratus. 'I hadn't thought of that.'

'The double whammy only works for big primes. Bigger than 50, which is half of 100.'

'Right. So I have to choose 2 now. Because if I chose 1, you'd choose 97 again. Or 89. And I'd lose.' So he chose 2. And eventually he lost. 'It's still a silly game,' he protested. 'I should have started with 97.'

'True. But you were the one who insisted on playing before I told you the *fourth* rule, which is designed to prevent double whammies.' And she wrote:

- The opening move must be an even number.

'Now it's a sensible game,' said Mathophila. And they played quite a long game, without much regard for tactics, which illustrates the rules nicely.

Move	Mathophila	Innumeratus	Comments
1	48		Even number, as required by rule 4
2		96	Doubles Mathophila's choice
3	32		One-third of Innumeratus's choice
4		64	Innumeratus is forced to choose a power of 2
5	16		So is Mathophila
6		80	Multiply by 5
7	10		Divide by 8
8		70	Multiply by 7
9	35		Halve
10		5	Only choices are 7 or 5 (or 1 and lose)
11	25		
12		75	Choices are 50, 75 and 100
13	3		
14		81	
15	9		Only 27 and 9 available
16		27	Bad move!
17	54		Forced since 1 loses
18		2	A better move is 18

19	62		Inspired variant on double whammy
20		31	Forced
21	93		Only choice, but a good one
22		1	Forced, and loses, because …
23	97		Effectively a double whammy – 93 is not prime but 31 has been used up

I suggest that at this point you stop reading, make a set of cards, and play the game for a while. I'm going to ask you to figure out a winning strategy, and it helps to have played the game. Anyway, it's a lot of fun.

Done that? Now we can get theoretical. Let's look at a simplified version where the cards run from 1 to 40. That will get you started.

Some opening moves lose very quickly. For example:

Move	Mathophila	Innumeratus	Comments
1	38		
2		19	Since 38 has gone, this is like a big prime
3	1		Forced
4		37	Mathophila loses

An opening move of 34 suffers the same fate.

Some numbers are best avoided altogether – like 1 in the 100-card game. Suppose Mathophila is silly enough to play 5. Then Innumeratus strikes back:

Move	Mathophila	Innumeratus	Comments
n	5		
$n+1$		25	
$n+2$	1		Forced and losing

Note that 25 *must* still be available when needed here, despite any previous moves, because it can be chosen only if the previous player plays 1 or 5.

There's a hint of a winning strategy here. Mathophila knows she's in trouble if she chooses 5, so she could try to force Innumeratus to choose 5 instead. Can she do this? Well, if Innumeratus chooses 7 then she can choose 35, and then Innumeratus has to choose 1 or 5, both of which lose.

Yes, but can she force Innumeratus to play 7? Well, if Innumeratus chooses 3 then Mathophila can choose 21, forcing Innumeratus to choose 7. Yes, but how does she make Innumeratus choose 3? Well, if he chooses 13, then Mathophila chooses 39 . . .

Mathophila can keep building hypothetical sequences of moves, all of which force Innumeratus's reply at every stage and lead to his inevitable defeat. The big question is: can she trap Innumeratus into such a sequence?

At some stage someone has to choose an even number, so we need to think about card 2. This is crucial because if Innumeratus chooses 2 then Mathophila can choose 26, forcing Innumeratus into the trap of playing 13. So now we come to the crunch: how can Mathophila force Innumeratus to choose 2?

She has to play an even number, and the more divisors this has, the more choices Innumeratus has, and Innumeratus might escape the trap. Anyway, the analysis gets complicated, too. Keep it simple. Suppose Mathophila opens with 22, twice a (smallish) prime. Then Innumeratus either chooses 2 and falls into Mathophila's trap – the long sequence of forced moves just outlined – or he chooses 11. If Mathophila plays 1 she loses, so she chooses 33 instead. Now 11 has already been used, so Innumeratus is forced to choose 3 – and the trap is sprung. We already know how Mathophila can win when he does that. So Mathophila *must* win if she starts with 22.

That's probably a bit confusing by now, so here's a summary of Mathophila's winning strategy. The two sets of columns deal with the two alternatives available to Innumeratus. For simplicity, I've assumed throughout that both players avoid 1, since it is an instant loss. With this choice eliminated, virtually every move is forced.

Move	Mathophila	Innumeratus	Mathophila	Innumeratus
1	22		22	
2		11		2
3	33		26	
4		3		13
5	21		39	
6		7		3
7	35		21	
8		5		7
9	25		35	
10		LOSE		5
11			25	
12				LOSE

There is at least one other opening move for Mathophila that also lets her force a win: if she chooses 26, then the same kind of game develops, but with a few of the moves interchanged.

The crucial features of Mathophila's strategy are the primes 11 and 13. Her opening move is twice such a prime: 22 or 26. It forces Innumeratus to reply either with 2 – at which point Mathophila is home and dry – or the prime. Then Mathophila replies with three times the prime, forcing Innumeratus to go to 3 – and she's home and dry again.

So Mathophila escapes trouble because as well as twice the prime, there is exactly one other multiple of such a prime in the range being played, namely 33 or 39. This provides her with an escape route. Call these the *medium* primes – they lie between one-third and one-quarter of the number of cards. If Mathophila chooses twice a medium prime, then Innumeratus must choose that prime. Then she chooses three times that prime, forcing Innumeratus to play the number 3.

Here are two questions for you:

- Can Mathophila win by any other strategy?
- Is there an analogous winning strategy for the 100-card version, and who wins?

More ambitiously, consider the game JG-n with the same rules, using an arbitrary whole number n of cards. Since no draws are allowed, and every game stops after finitely many moves, game theory implies that either Mathophila has a winning strategy, or Innumeratus does.

- With perfect strategy, who wins JG-n, assuming Mathophila goes first?

Certainly the answer depends on n. Mathophila wins when n is 3 or 8, whereas Innumeratus wins when $n = 1$, 2, 4, 5, 6, 7, 9. What about $n = 100$? What about all the values of n from 10 to 99? Can you solve the whole thing?

Answers on page 333

• •

Mathematical Metajoke

An engineer, a physicist, and a mathematician found themselves in a joke, very similar to many that you will have heard before, but did not immediately realise where they were.* After a hasty back-of-the-envelope calculation, the engineer worked out what had happened and began to chuckle. Soon after, the physicist intuited where they were, based on a loose analogy with a particle confined in a box, and began laughing uproariously. The mathematician, however, seemed not to find their situation remotely funny. Eventually the others asked why.

'I saw immediately that I was in an anecdote of some kind,' he replied. 'But it was only after I noticed characteristic structural features that I could be sure the anecdote was a joke. However, this joke is far too trivial a consequence of the general case to have any amusement value.'

• •

---

* They thought it was a bar.

Beyond the Fourth Dimension

Physicists are seeking a Theory of Everything (ToE) that will unify the two pillars of modern physics, relativity and quantum mechanics, while fixing certain inconsistencies between these two theories. The search has led them to speculate that our familiar 3-dimensional (3D) space is not actually 3D at all, but 10D or maybe 11D. The extra dimensions provide a place for fundamental particles to vibrate in (like a violin string), thereby giving rise to quantum numbers such as spin and charge (which are like the notes produced by the violin string). Now, you might think that it would be difficult for everyone to have been so wrong for so long about something so basic as the dimensionality of space. And in any case, surely space is space, and it can't have 10 dimensions because there isn't any room to fit 7 more in once we've sorted out the first 3.

However, it's not that simple. Mathematicians have invented logically consistent geometries with 4, 5, 6, or even infinitely many, dimensions. Whichever number you like, including 10. So, on the face of it, there is nothing sacred about 3D space. It might be a historical accident, which could have been different in another run of the universe. It might be sacred after all, the only possibility for reasons we don't yet appreciate. Or it might not actually *be* 3D, despite appearances. And even if it is, there's no reason to expect it to be the neat, tidy 3D space of Euclid. In fact, thanks to Einstein's general theory of relativity, we think space is curved, in ways that Euclid never dreamed of, and sort of mixed together with bits of time.

A century and a half ago, Victorian England faced a similar problem, the equally baffling concept of the Fourth Dimension. Mathematicians had encountered it when looking for something else: William Rowan Hamilton had spent decades searching for a natural algebra of 3D space, much as complex numbers are a natural algebra of 2D space, and had been forced instead to settle for a natural algebra of 4D space, which he called quaternions. Scientists were finding that 4D thinking helped them sort out a lot of basic physics. Spiritualists, claiming to put people in contact with the dead, realised that the Fourth Dimension was a great place to locate the Spirit World, because we can't go there to check whether spirits exist, but they can see us from their superior vantage point if they do. The same went for ghosts, another Victorian obsession; ghosts could appear from, and disappear into, that extra dimension. And what we now refer to as 'hyperspace theologians' were quick to see the advantages of locating God in the Fourth Dimension. From there, He could watch over every part of His creation, while remaining outside it – just as we can see an entire printed page at a glance, without being embedded in the paper.

All this interest was somewhat tempered by a contrary view: the Fourth Dimension does not exist, indeed *cannot* exist: it is inconceivable. The debate managed to muddle two distinct questions: the structure of physical space, and the possibility of logically consistent mathematical spaces that differ from the orthodox 3D model. Philosophers got in on the act, mostly arguing for the orthodox 3D model – which was a bit surprising given their propensity to claim that nothing really exists and everything we perceive is an illusion.

Into this intellectual hornet's nest dropped a polite clergyman and headmaster of a prominent boys' school, Edwin Abbott Abbott. Yes, two Abbotts, to distinguish him from his father, Edwin Abbott. In 1884 Abbott published one of the most curious and original books ever written, a mathematical fantasy called *Flatland*.

Edwin Abbott Abbott ...   ... and his book.

Abbott hit on a cunning way to get his fellow Victorians to accept the possibility of a Fourth Dimension, by quietly pursuing an analogy in which 2D creatures living in a 2D world found the very idea of 3D space inconceivable, indeed heretical – and only then springing 4D on his readers, when he had softened them up. His hero, the modestly named A. Square,* lives a humdrum life with his linear wife and polygonal children in a pentagonal house in the 2D universe of Flatland, which is a Euclidean plane. Abbott wove in some biting satire about the suppression of women and the poor in Victorian society, as well. And a few quotations from Shakespeare, and some allusions to Aristotle.

Anyway, A. Square lives in a 2D universe and can conceive of no other. That is how things are, that is how things have always been, and that is how things always will be. And in any case, the Third Dimension is religious heresy and the priesthood of Circles will make short shrift of anyone who dares to mention it. And so A. Square continues his humdrum life, until one day he undergoes an epiphany, and becomes totally converted to the notion of a 3D world. This change of heart is triggered by a visitation from ... the Sphere.

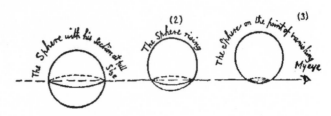

A. Square meets the Sphere.

Now, the limitations of his 2D nature prevent him from seeing the Sphere as a single object. Instead, he sees the circles† in which it meets his flat world. A dot materialises into thin air in an empty room. Then it grows to a circle, expands to a larger

---

* Abbott Abbott $= A^2$?
† He also sees the circles 'edge on'. Just as we see only a 2D projection – or a stereo pair of projections – of a 3D object.

circle, then shrinks back to a dot and vanishes. (You see why Victorian ghostbusters liked the idea of 4D.) He thinks the Sphere is some sort of priest, but one capable of changing its size. We Spacelanders have the luxury of visualising the geometry: a Sphere of fixed size passes through the Flatland plane, and the intersection changes as it does so.

Now, says Abbott – though not in these words – a Victorian Spacelander, trying to contemplate the Fourth Dimension, is in the same position as A. Square trying to contemplate the Third. Protestations about the natural order, or the alleged impossibility of extra dimensions, carry no more weight in Spaceland than they do in Flatland. Abbott confines his mathematical discussion to enumerating the edges and vertices of a cube, and a 4D hypercube, but the point comes over quite strongly.

By analogy, if we Spacelanders met a Hypersphere from the Fourth Dimension, we would perceive only the sequence of spheres that arises when it meets our 3D space. Like A. Square, we see a dot, materialising into thin air in an empty room. Then it grows to a sphere, expands to a larger sphere, then shrinks back to a dot and vanishes. (You see even more clearly why Victorian ghostbusters liked the idea of 4D.)

We can turn this loose geometric analogy into solid algebra, using coordinates. We are used to expressing a point in the plane using two numbers (x, y). Similarly, points in space can be expressed as triples (x, y, z). Here our familiar 3D space runs out of new directions, but mathematically we can explore the behaviour of quadruples (x, y, z, w), and this is what mathematicians mean by a 4D space. Such a space comprises *all* possible quadruples, not just one. And it has a natural 'geometry', because we can define distances using an extension of Pythagoras's theorem, and once we have distances, we also have angles, and circles, and most of the other stuff that we associate with geometry. Now we can say what we mean by hyperspheres, hypercubes, and all sorts of cute geometrical objects. It all fits together beautifully, and once we've become accustomed to the language, these new types of space start to feel just as real as the one we live in.

Around 1900, physicists and mathematicians suddenly realised the advantages of thinking of time as a (not 'the') Fourth Dimension. Soon everyone was very happy talking about 4D spacetime. Today, designers of video games talk of 4D graphics, meaning 3D graphics that *move*. If we view A. Square's encounter with the Sphere as a movie, we are in effect using time as a surrogate for a third spatial dimension. Our own encounter with a hypersphere can be visualised using time as a surrogate for a fourth spatial dimension.

However, that's a surrogate, not the reality. The Sphere from the Third Dimension existed, unchanged, as time passed. Only his intersection with Flatland changed in time. Moreover, time is not the only surrogate for an extra dimension of space. We could employ colour, or temperature, or an entirely new physical quantity, instead.

For instance, suppose the dimension of 'colour' varies from yellow to blue through intermediate shades of green, and the universe is a plane on which coloured figures move about. By a trick of perception, they interact, and perceive each other, only when they have the same colour. Now the green creatures would mimic Flatland. So would the yellow ones, and the blue ones. But these three 'parallel worlds' really are parallel – in the sense of not meeting. They are separated along the 'colour dimension'. Now a Sphere could be represented as a yellow dot surrounded by circles that become greener as they expand, and then shrink back to a blue dot. From our 3D perspective we could pull them apart 'along the colour dimension' and see the whole thing as a conventional geometric sphere, with shades of colour parallel to its equator. But we don't have to do that: the colour image is entirely adequate.

When we call the set of quadruples of numbers a *space*, we are emphasising the 4D analogues of traditional 3D geometry. However, the numbers that appear in the quadruple (x, y, z, w) do not have to be spatial measurements in the usual sense. They might, for example, be coordinates

(price, colour, weight, temperature)

in the space of all woolly pullovers, with colour ranging on a numerical scale from yellow (0) to blue (1). A specific woolly pullover, with coordinates

(27.43, 0.62,1.37, 22.61)

would have

price = £27.43
colour = bluish-green
weight = 1.37 kilograms
temperature = 22.61°C

So, although a woolly pullover is a 3D object, we are representing a few of its key features in a 4D mathematical space. In short, woolly-pullover-space is 4D.

Economists use this approach to represent the state of the nation's economy, but now they work in, say, a space with a million dimensions, whose coordinates show the prices of a million goods. Astronomers represent the locations and velocities of the eight planets of the solar system[*] using six numbers for each planet – three for location and three more for velocity. So the state of the planets, at any given time, defines a point in a 48-dimensional space.

Just as A. Square discovered – to his initial incredulity – that his 2D world was really just part of a higher-dimensional universe, so physicists are beginning to wonder whether the same applies to our 3D world. According to string theory – well, one popular version of many different string theories – space may actually be 10D. The number 10 is not an arbitrary choice, but because this kind of ToE works only in 10 dimensions.

Of course, string theory may not correspond to reality. But

---

[*]   Alas! Poor Pluto.

science has taught us many times that the world is more complicated than the one we perceive. If relativity and quantum theory are ever unified, then our view of our world will have to change, just as it did when those two theories were first proposed.

All very well, but: Why don't we notice those missing dimensions?

There are at least three possible answers.

- They don't exist, and string theory is wrong.
- They do exist, but they are curled up so small that we can't see them. From a distance, a hosepipe looks 1D, but close up it has a circular cross-section, adding two more dimensions. If that cross-section were really really small – much smaller than the diameter of an electron, say – the hosepipe would be convincingly 1D unless you developed very delicate experimental techniques to probe those hidden dimensions. Now replace the hosepipe by our apparently 3D space, and the circular cross-section by an equally tiny 7D sphere, and you'll get the idea.
- Our space really is 3D, but it is embedded in a surrounding 10D space – and we don't perceive the bigger space because we can't look, or move, in those directions. Just as A. Square was confined to the plane of Flatland, we may be confined to a 3D slice of that 10D space. Mathematically, this kind of behaviour is entirely natural: dynamical systems often have 'invariant subspaces', and anything that lives in those subspaces can't escape from them. Try moving yourself into the past to appreciate what I mean. Physicists have taken to calling such subspaces 'branes', a term derived from 'membrane' via 'm-brane', an m-dimensional subspace.

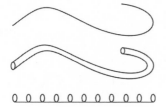

A hosepipe seems to be 1D, but closer up we see it has two more dimensions. We can draw this schematically as a line with circles attached to each point.

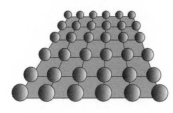

Extra dimensions of space (here the 2D plane) shown schematically as spheres. In string theory, the spheres have more dimensions than we can draw. The spheres support quantum vibrations, which endow particles with properties like spin and charge.

All this talk of 'hidden dimensions' may be needlessly mystical. Physics presented us with very similar things long ago, but no one started babbling about increasing the dimension of space. An electromagnetic field – which we use to send radio, TV and phone calls – has six extra coordinates for each point in space: three for the magnetic field's strength and direction, and three more for the electric field's strength and direction. Maxwell's equations for electromagnetism are naturally defined on a 9D space.

So the extra 7 dimensions required for string theory need not actually be *spatial* in any meaningful sense. They might be – in fact, *are* – new physical quantities, like colour or temperature, that enter into the string theory equations. So talking of them as hidden dimensions of *space* makes string theory seem more mysterious than it really is.

Slade's Braid

In the 1880s, the American medium Henry Slade used to convince people that he had access to the Fourth Dimension – the Spirit World – using a strip of leather with two cuts along it. He would get someone to make a mark on the leather, to prevent substitution. Then he would hold it under the table for a few moments, and produce it again – braided!

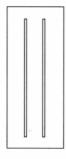 

Start here ...   ... end here.

In 4D space, strips can be passed over each other and woven together by pushing one temporarily into the Fourth Dimension, moving it into the right position, and then pulling it back into ordinary 3D space. That is what Slade pointed out, and what he claimed proved he had the ability to access the Fourth Dimension.

How did he do it?

Answer on page 334

• •

Avoiding the Neighbours

Place each of the digits 1–8 in the eight circles, so that neighbouring digits (that is, those that differ by 1) do not lie in neighbouring circles (connected directly by a line).

Answer on page 335

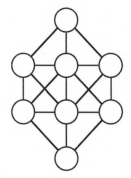

Keep the neighbours apart.

Career Move

A mathematician who had spent his entire research career in pure mathematics – starting with topological algebra, then a bit of algebraic geometry, then some geometric topology, thinking of moving into algebraic topology or maybe geometric algebra – began to wonder if perhaps it was time he did something more obviously practical. He knew that those subjects did have applications, but he had never worked on such things, preferring the intellectual challenges of abstract thought.

He had never been *against* applied mathematics, you understand – just hadn't done any himself.

Maybe, he thought, *it's time for a change.*

Weeks went by, and still he had not translated his thoughts into deeds. The prospect of engaging with the real world made him very nervous. He'd never done it before. But he found the idea appealing, nonetheless. The problem was to pluck up enough courage to take the plunge.

One day, walking along the corridor of the Mathematics Department, he saw a sign on a door. 'Seminar on gears – today at 2.00.' He looked at his watch: 1.56. Dare he? Could he actually . . . *go in*? It was a big step. In an agony of indecision, he stood outside the door, shifting from one foot to the other, listening to the sounds of the lecturer preparing to start the talk. Finally, at 1.59, he plucked up his courage, opened the door, and slid into a vacant seat. Now he would begin his career move to practical applications of mathematics!

The speaker picked up his notes, cleared his throat, and began. 'The theory of gears with an *integer* number of teeth is well known—'

A Rolling Wheel Gathers No Speed

A wheel of radius 1 metre rolls along a flat horizontal road at a constant speed of 10 metres per second, without slipping and without bouncing off the road. At a fixed instant of time, is any point on the wheel stationary? If so, which?

Assume that the wheel is a circular disc, the road is a straight line, and the wheel lies in a vertical plane. 'Stationary' means that the instantaneous velocity is zero.

Answer on page 335

Point Placement Problem

You have a line of unit length, whose two endpoints at 0 and 1 are missing, and an unbounded supply of points – as one does. You are required to place the points successively on the line, so that:

- The second point and the first point lie in different halves of the line. (To avoid ambiguity, the midpoint at $\frac{1}{2}$ is excluded: neither point is allowed to lie in that exact position. So one 'half' runs from 0 to $\frac{1}{2}$, excluding both, and the other runs from $\frac{1}{2}$ to 1, excluding both.)
- The third point and the first and second points all lie in different thirds of the line. (To avoid ambiguity, the points at $\frac{1}{3}$ and $\frac{2}{3}$ are now excluded.)
- The fourth point and the first, second and third points all lie in different quarters of the line. (The points at $\frac{1}{4}$ and $\frac{3}{4}$ are now excluded – remember, we have already excluded $\frac{1}{2}$.)

Now keep going, obeying, for increasing n, this rule:

- The nth point and the previous $n - 1$ points all lie in different $\frac{1}{n}$ ths of the line. (All points $\frac{m}{n}$, for m = 0, 1, 2, ..., n, are excluded.)

Got that? Here's the question: how long can you keep this process going?

At first sight, the answer seems to be: as long as you wish. After all, you can divide the line into indefinitely ever finer pieces, and choose points in whichever of those are appropriate.

I really don't expect you to get the correct answer here, but I don't want to give it away immediately, so you'll find it on page 336. It's amusing to try placing the first five or six points. Even then, it's not as easy as it sounds.

• •

Chess in Flatland

In *Flatland* (page 255) the world is a plane and its inhabitants are geometric shapes. Flatlanders play their own versions of Spaceland games, and one of them is chess. The Flatland chessboard is eight cells long, and each player has three pieces: king, knight and rook, starting in the position shown.

Start of a game of Flatland chess.

The rules resemble those of Spaceland chess, bearing in mind the limitations of Flatland geometry. All three pieces can move to the left or to the right, if a suitable space is available. All moves must end either on an empty cell, or on a cell occupied by an enemy piece, which is then removed from the board – 'taken'.

- A *king* (the piece with a cross on top) moves only one cell at a time, and cannot move into 'check' – a cell that is already threatened by the enemy.
- A *knight* (horse shape) moves by jumping over an adjacent cell, which may be empty or occupied, and landing on the one on its far side. So it ends up two cells away from where it started.
- A *rook* (castle shape) can move across any number of unoccupied cells.

If a player has no legal move available, the game is stalemate, and is a draw. If a player can threaten the opposing king, and the king cannot escape, that's checkmate, and the game is won.

If White plays first, and both players adopt perfect strategy, who will win?

Answer on page 337

Answer on page 337

The Infinite Lottery

The Infinite Lottery involves infinitely many bags: one numbered 1, one numbered 2, one numbered 3, one numbered 4, and so on. Each bag contains infinitely many lottery balls with the corresponding number.

You are supplied with a large box. You may place any number of balls you like into this box, chosen from any of the bags. There is just one condition: the total number must be finite.

Now you are required to change the balls in the box. You must remove and discard one, and replace it by as many balls as you like that bear smaller numbers. For instance, if you discard a ball with the number 100 on it, you can add to the box 10 million balls with the number 99, 17 billion with the number 98, and so on. There is thus no upper limit to the number of balls that can replace that solitary number-100 ball.

You must keep doing this. At each stage you may replace the discarded ball by whichever combination of balls you wish, provided you make their number finite and make sure that they bear smaller numbers than the ball you discarded. If you remove a ball marked 1, you cannot replace it because there are no balls with smaller numbers on them.

If eventually you run out of balls and empty the box, you lose. If you keep removing balls for ever – that is, if you never run out of balls – then you win.

But *can* you win the infinite lottery? If so, how?

Answer on page 337

Answer on page 337

●●●

Ships That Pass ...

In the days when people crossed the Atlantic in passenger liners, a ship left London every day at 4.00 p.m. bound for New York, arriving exactly 7 days later.

Every day at the same instant (11.00 a.m. because of the time difference) a ship left New York bound for London, arriving exactly 7 days later.

All ships followed the same route, deviating slightly to avoid collisions when they met.

How many ships from London does each ship sailing from New York encounter during its transatlantic voyage, *not* counting any that arrive at the dock just as they leave, or leave the dock just as they arrive?

Answer on page 339

• •

The Largest Number Is Forty-Two

Mathematicians often use a technique called proof by contradiction. The idea is that to prove some statement true, you begin by assuming it to be false, and go on to derive various logical consequences. If any of these consequences leads to a logical impossibility – a contradiction – then your assumption that the statement is false cannot be correct. Therefore the statement is true.

You may have come across this by the name used in Euclid, which in Latin translation is *reductio ad absurdum* – reduction to the absurd.

For example, to prove that pigs don't have wings, you first assume that they do, and deduce that pigs can fly. But we know that they can't, so this is a logical impossibility. Therefore it is false that pigs don't have wings, so they do.

Got that?

I will now use proof by contradiction to show that the largest whole number is 42.

Let n be the largest whole number, and suppose for a contradiction that n is not 42. Then $n > 42$, so $(n - 42)^3 > 0$, which expands to give

$$n^3 - 126n^2 + 5{,}292n - 74{,}088 > 0$$

Adding n to each side,

$$n^3 - 126n^2 + 5{,}293n - 74{,}088 > n$$

But the left-hand side is a whole number. Since it is greater than n, which we are assuming to be the *largest* whole number, we have derived a contradiction.

Therefore it is false that the largest whole number is not 42. So it *is* 42!

Clearly something is wrong here – but what?

Answer on page 339

• •

A Future History of Mathematics

2087   Fermat's Lost Theorem is found again on the back of an old hymn-sheet in the Vatican secret archives.

2132   A general definition of 'life' is formulated at the Intercontinental Congress of Biomathematicians.

2133   Kashin and Chypsz prove that life cannot exist.

2156   Cheesburger and Fries prove that at least one of Euler's constant, the Feigenbaum number, and the fractal dimension of the universe is irrational.

2222   The consistency of mathematics is established – it is that of cold sago pudding.

2237   Marqès and Spinoza prove that the undecidability of the undecidability of the undecidability of the undecidability of the P=NP? problem is undecidable.

2238   Pyotr-Jane Dumczyk disproves the Riemann hypothesis by showing that there exist at least 42 zeros $\sigma + it$ of the zeta function with $\sigma \neq \frac{1}{2}$ and $t < \exp\exp\exp\exp\exp((\pi^e + e^\pi)\log 42)$.

2240   Fermat's Lost Theorem is lost again.

2241	Sausage conjecture proved in all dimensions except 5, with the possible exception of the 14-dimensional case, where the proof remains controversial since it seems too easy.
2299	Contact is made with aliens from Grumpius, whose mathematics includes a complete classification of all possible topologies for turbulent flows, but has been stuck for the past five galactic revolutions because of an inability to solve the $1 + 1 = ?$ problem.
2299	The solution of the $1 + 1 = ?$ problem by Martha Snodgrass, a six-year-old schoolgirl from Woking, ushers in a new age of Terran–Grumpian cooperation.
2300	Formulation of Dilbert's 744 problems at the Interstellar Congress of Mathematicians.
2301	Grumpians depart, citing the start of the cricket season.
2408	Riculus Fergle uses Grumpian orthocalculus to show that all of Dilbert's problems are equivalent to each other, thereby reducing the whole of mathematics to a single short formula.*
2417	The DNA-superstring computer Vast Intellect fails the Turing Test on a technicality but declares itself intelligent anyway.
2417	Vaster Intellect invents the technique of human-assisted proof, and uses it to prove Fergle's Final Formula, with the Dilbert problems as corollaries.
2417	Even Vaster Intellect discovers inconsistencies in the operating system of the human brain, and all human-assisted proofs are declared invalid.
7999	Grunt Snortsen invents counting on his toes; Reign of the Machines comes to an abrupt end.†

---

* The famous $\in \mho \mathfrak{M}\mathbb{C}^{42}$. Plus a constant.

† Snortsen had lost a toe in an encounter with a berserk cash-register.

11,868	The Rediscovery of Mathematics, now in base 9.
0	Reformation of the calendar.
1302*	Fergle's Final Formula is proved, correctly this time, and mathematics *stops*.
1302†	Diculus Snergle asks what would happen if you allowed the arbitrary constant in Fergle's Final Formula to be a variable, and mathematics starts up again.

* *

---

* 17 May, 2.46 p.m.
† 17 May, 2.47 p.m.

Professor Stewart's Superlative Storehouse of Sneaky Solutions and Stimulating Supplements

Wherein the perspicacious or perplexed reader may procure answers to those questions that are presently known to possess answers ... together with such gratuitous facts and fancies as may facilitate their further delectation and enlightenment.

Calculator Curiosity 1

$(8 \times 8) + 13 = 77$

$(8 \times 88) + 13 = 717$

$(8 \times 888) + 13 = 7117$

$(8 \times 8888) + 13 = 71117$

$(8 \times 88888) + 13 = 711117$

$(8 \times 888888) + 13 = 7111117$

$(8 \times 8888888) + 13 = 71111117$

$(8 \times 88888888) + 13 = 711111117$

• •

Year Turned Upside Down

Past: 1961; *Future*: 6009.

If you insist on allowing a squiggle on the 7, amend these to 2007 and 2117.

• •

Sixteen Matches

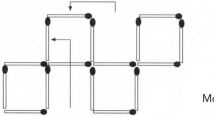

Move these two.

• •

Swallowing Elephants

The deduction is false.

Suppose, for the sake of argument, that elephants are easy to swallow. Then the third statement in the puzzle tells us that elephants eat honey. The second then tells us that elephants can play the bagpipes. On the other hand, the first statement tells us that elephants wear pink trousers, in which case the fourth statement tells us that elephants *can't* play the bagpipes. So we get a logical contradiction. The only way out is if elephants are not easy to swallow.

There's a systematic method for answering such questions. First, turn everything into symbols. Let

E be the statement: 'Is an elephant.'
H be the statement: 'Eats honey.'
S be the statement: 'Is easy to swallow.'
P be the statement: 'Wears pink trousers.'
B be the statement: 'Can play the bagpipes.'

We use the logical symbols

$\Rightarrow$ meaning 'implies'
$\neg$ meaning 'not'.

Then the first four statements read:

$E \Rightarrow P$
$H \Rightarrow B$
$S \Rightarrow H$
$P \Rightarrow \neg B$

We need two of the mathematical laws of logic:

$X \Rightarrow Y$ is the same as $\neg Y \Rightarrow \neg X$
If $X \Rightarrow Y \Rightarrow Z$, then $X \Rightarrow Z$

Using these, we can rewrite the conditions as:

$E \Rightarrow P \Rightarrow \neg B \Rightarrow \neg H \Rightarrow \neg S$

so $E \Rightarrow \neg S$. That is, elephants are not easy to swallow.

This list of attributes suggests yet another way to get the answer: think about an elephant (E) that (P) wears pink trousers, ($\neg$ B) does not play the bagpipes, ($\neg$ H) does not eat honey, and ($\neg$ S) is not easy to swallow. Then all four statemnts in the puzzle are true, but 'elephants are easy to swallow' is false.

• •

Magic Circle

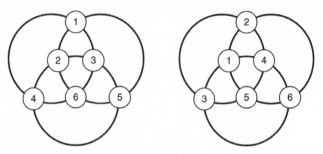

These or their rotations and reflections.

● ●

Press-the-Digit-ation

The explanation of Whodunni's calculator trick uses a bit of algebra.

Suppose you live in house number x, were born in year y, and have had z birthdays so far this year, which is either 0 or 1, depending on dates. Then successive steps in the trick go like this:

- Enter your house number: x
- Double it: $2x$
- Add 42: $2x + 42$
- Multiply by 50: $50(2x + 42) = 100x + 2100$
- Subtract the year of your birth: $100x + 2100 - y$
- Subtract 50: $100x + 2050 - y$
- Add the number of birthdays you have had this year: $100x + 2050 - y + z$
- Subtract 42: $100x + 2008 - y + z$

If we're doing the trick in 2009, then $2008 - y$ is one less than the number of years that have passed since your birth year. Adding the number of birthdays you've had this year leaves it that way if you haven't had one yet, but adds 1 if you have. The result is always your age. (Think about it. If you were born one year ago but haven't had a birthday yet, your age is 0. After your birthday, it's 1.)

So the final result is $100x +$ your age. So provided you are aged between 1 and 99, the last two digits will be your age (written as 01–09 if your age is 1–9). Removing those and dividing by 100, which is the same as looking at the remaining digits, gives x – your house number.

If you're over 99, then the final two digits can't be your age. There will be an extra digit (which barring medical miracles will be 1). So your age will be 1 followed by the final two digits. And your house number will be the rest of the digits except those two, *minus one*.

If you're age 0, the trick still works provided you count the day of your birth as a birth-day. Your zeroth, in fact. But usually we don't do that, which is why I excluded age 0.

To modify the trick for any other year, say $2009 + a$, just change the final step to 'subtract $42 - a$'. So in 2010 subtract 41, in 2011 subtract 40, and so on. If you're reading this after 2051, make that 'add $a - 42$'. It's the same thing, but it will sound more sensible.

• •

Secrets of the Abacus

To subtract (say) a 3-digit number $[x][y][z]$, which is really $100x + 10y + z$, we have to form the complement $[10 - x][10 - y]$ $[10 - z]$, which is really $100(10 - x) + 10(10 - y) + (10 - z)$. This is equal to $1000 - 100x + 100 - 10y + 10 - z$, or $1110 - (100x + 10y + z)$. So adding the complement is the same as subtracting the original number, but adding 1110. To get rid of it, subtract 1 from positions 4, 3, 2, but not 1.

• •

Redbeard's Treasure

Redbeard will locate the lost loot 128 paces north of the rock.

At each step the piratical finger can move either left or right – two choices. So the number of routes down the diagram doubles for each extra row. There are 8 rows, and only one T to start from, so the number of routes is $1 \times 2 \times 2 \times 2 \times 2 \times 2 \times 2 \times 2 = 128$.

If we replace each letter by the number of routes that lead to it, we get a famous mathematical gadget, Pascal's triangle:

$$
\begin{array}{ccccccccccccccccc}
 & & & & & & & & 1 & & & & & & & & \\
 & & & & & & & 1 & & 1 & & & & & & & \\
 & & & & & & 1 & & 2 & & 1 & & & & & & \\
 & & & & & 1 & & 3 & & 3 & & 1 & & & & & \\
 & & & & 1 & & 4 & & 6 & & 4 & & 1 & & & & \\
 & & & 1 & & 5 & & 10 & & 10 & & 5 & & 1 & & & \\
 & & 1 & & 6 & & 15 & & 20 & & 15 & & 6 & & 1 & & \\
 & 1 & & 7 & & 21 & & 35 & & 35 & & 21 & & 7 & & 1 & \\
1 & & 8 & & 28 & & 56 & & 70 & & 56 & & 28 & & 8 & & 1
\end{array}
$$

Here each number is the sum of the two above it to left and right, except down the sides where they're all 1's. If you add up the rows, you get the powers of two: 1, 2, 4, 8, 16, 32, 64, 128. So this is another – closely related – way to get the same answer.

• •

Stars and Snips

Fold the paper in half (say along the vertical line in my picture) and then fold it alternately up and down along the other lines to make a zigzag shape, the way a fan folds. Then cut along a suitable slanting line – and unfold. I've drawn the ghost of the star to show how it relates to the folds.

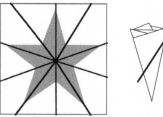

Fold ...   ... and cut.

Yes, you *can* make a six-pointed star in a similar way. If anything, it's easier: first fold the paper in quarters, then fold the result along two lines trisecting the right-angled corner. As for the five-pointed star, you have to snip at the correct angle. That's committees for you.

• •

The Collatz–Syracuse–Ulam Problem

The cycles that appear with zero or negative numbers are:

- $0 \to 0$
- $-1 \to -2 \to -1$
- $-5 \to -14 \to -7 \to -20 \to -10 \to -5$
- $-17 \to -50 \to -25 \to -74 \to -37 \to -110 \to -55 \to -164 \to -82$
  $\to -41 \to -122 \to -61 \to -182 \to -91 \to -272 \to -136 \to -68 \to$
  $-34 \to -17$

The Jeweller's Dilemma

The lengths contain 8, 7, 6, 6, 5, 5, 5, 4 and 3 links. Instead of breaking up one link on each chain, we could break all 8 links in the 8-link piece, and use these to join the other eight pieces together: total cost £24. But there's a cheaper way. Break the pieces of lengths 4 and 3 into separate links, and use these to join the seven other pieces. The total cost is now £21.

What Seamus Didn't Know

No, it doesn't wave its paws madly and exploit air resistance to create a force, like the wings of a bird do. Instead, the cat manages to change its orientation without causing any change to its angular momentum at any time.

- *Initial position*: cat upside down, stationary, angular momentum zero.
- *Final position*: cat right side up, stationary, angular momentum zero.

No contradiction there, but of course there's the bit in between, when the cat starts to rotate. Except – it doesn't. Rotate, that is. A cat is not a rigid body.* In 1894, the French doctor Étienne Jules Marey took a series of photographs of a falling cat.

---

* Except when you're trying to stuff it into the cat basket, to take it to the vet.

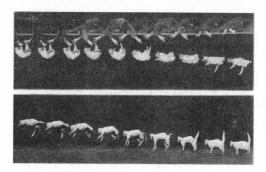

Marey's cat experiment.

The secret was then revealed. Because a cat is not a rigid body, it does not have to rotate its entire body *simultaneously*. Here is the cat's recipe for turning over, while maintaining zero angular momentum throughout:

- Pull in your front legs and spread out your back legs.
- Twist your front half quickly one way, and your back half slowly the other way. Your two halves move with opposite angular momentum, so the total remains zero.
- Spread out your front legs and pull in your back legs.
- Turn your back half quickly to align with the front half, while your front half turns slowly backward. Again your two halves move with opposite angular momentum, so the total remains zero.
- Your tail can also move, and usually does, assisting the process by providing a useful reservoir of spare angular momentum.

Modern photo of falling cat.

Lincoln's Dog

Well, the dog may have lost its tail, or some legs, or it might be a mutant with six legs and five tails ... Quibbles aside, this is a good question to distinguish mathematicians from politicians. Lincoln asked his question in the context of slavery, putting it to a political opponent who maintained that slavery was a form of protection, trying to imply that it was benign. Lincoln's answer was: 'It has four legs – calling a tail a leg doesn't make it a leg.' By which he meant that calling slavery protection didn't *make* it protection, which in that context is fair enough. Barack Obama's famous 'lipstick on a pig' remark made the same point, though his opponents chose to interpret it as an insult to Sarah Palin.*

Ignoring the context, though, most mathematicians would beg to differ with President Lincoln, and answer 'five'. Renaming a tail as a leg amounts to a temporary redefinition of terminology, which is common in mathematics. For example, in algebra 'the' unknown is usually denoted x, but the value of x differs from one problem to the next. If x was 17 in last week's homework it need not be 17 ever after. The usual convention is that a temporary redefinition *remains in force* until it is explicitly cancelled, or until the context makes clear that it has been cancelled.

In fact, mathematicians habitually go further, and permanently redefine important terminology, usually to make it more general. Concepts such as number, geometry, space and dimension are examples; their meaning has changed repeatedly as the subject has progressed.

So, to mathematicians, if we agree to call a tail a leg for the rest of the discussion – which Lincoln's question tacitly assumes, otherwise it's not worth asking – then the meaning of 'leg' has *changed*, and now includes tails. So, Mr President, the dog has *five* legs, by your own redefinition.

---

* This was unfair to pigs, and ignored a long tradition of political piggery, including the book *Lipstick on a Pig: Winning in the No-Spin Era by Someone Who Knows the Game* by Victoria Clarke, an Assistant Secretary in George W. Bush's administration. See: en.wikipedia.org/wiki/Lipstick_on_a_pig.

What happens to Lincoln's political point? It remains intact, but for a different reason. When Lincoln's opponent asserted that slavery is protection, he redefined protection for the remainder of the discussion, so the properties normally associated with protection might not apply any longer. In particular, the new meaning does not imply that slavery is an act of kindness.

. . . .

Whodunni's Dice

The dice were 5, 1 and 3.

If the dice show the numbers a, b and c, then the calculation produces in turn the numbers

$$2a + 5$$
$$5(2a + 5) + b = 10a + b + 25$$
$$10(10a + b + 25) + c = 100a + 10b + c + 250$$

So Whodunni subtracted 250 from 763, to get 513 – the numbers on the three dice. Subtract 2 from the first digit of the answer, 5 from the second, and leave the third alone – easy.

. . . .

The Bellows Conjecture

Heron's formula applies to a triangle with sides a, b, c, and area x. Let s be half the perimeter:

$$s = \tfrac{1}{2}(a + b + c)$$

Then Heron proved that

$$x = \sqrt{s(s - a)(s - b)(s - c)}$$

Square this equation and rearrange to get

$$16x^2 + a^4 + b^4 + c^4 - 2a^2b^2 - 2a^2c^2 - 2b^2c^2 = 0$$

This is a polynomial equation relating the area x to the three sides a, b, c.

. . . .

Digital Cubes

The other 3-digit numbers that are equal to the sum of the cubes of their digits are 370, 371, and 407.

If the digits are *a*, *b*, *c,* then we have to solve

$$100a + 10b + c = a^3 + b^3 + c^3$$

with $0 \leq a, b, c \leq 9$ and $a > 0$. That's 900 possibilities, so a systematic search will find the answer.

The work can be reduced by using some fairly simple tricks. For instance, if you divide a perfect cube by 9, the remainder is 0, 1 or 8. If you divide 100 or 10 by 9, the remainder is 1. So $a + b + c$ and $a^3 + b^3 + c^3$ leave the same remainder on division by 9. Eliminating cases where the digits are too small or too big to work, $a + b + c$ has to be one of 7, 8, 9, 10, 11, 16, 17, 18, 19, 20. After that … well, you get the idea. It's a bit of a scramble, but it can be pushed through. Maybe there's a better way.

● ◦

Order into Chaos

There are plenty of solutions (usually there are lots, or none). Here's one for each puzzle:

- SHIP–SHOP–SHOT–SOOT–ROOT–ROOK–ROCK–DOCK
- ORDER–OLDER–ELDER–EIDER–CIDER–CODER–CODES–CORES–SORES–SORTS–SOOTS–SPOTS–SHOTS–SHOPS–SHIPS–CHIPS–CHAPS–CHAOS

If you're worried about EIDER and SOOTS, the first is a kind of duck and the second is not the plural of 'soot' (which is 'soot') but derives from the verb 'to soot', which means to cover with soot. As in the lesser-known proverb 'A clumsy chimney-sweep soots the hearth.'* Both words are in the official Scrabble™ dictionary.

Now, I promised some maths, and we've not seen any yet.

All these puzzles are really about networks (also called graphs), which are collections of dots joined by lines. The dots represent

---

* I made this up, but it wouldn't surprise me to find it's been used for centuries in parts of Lincolnshire.

objects, and the lines are connections between these. In the SHIP–
DOCK puzzle, the objects are four-letter words, and two words are
connected if they differ by just one letter (in a specific position). All
four-letter word puzzles of this kind reduce to the same general
question: Is the initial word connected to the final one by some path
in the network of all possible four-letter words?

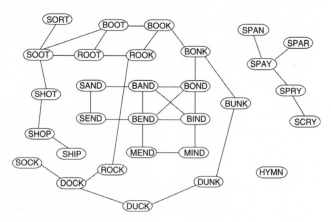

Connect SHIP to DOCK.

The diagram shows just a tiny part of this network – enough to
find an answer.

There are computer algorithms (procedures) for finding paths
between any two nodes of a network, and the mathematics quickly
becomes fairly deep and difficult. One relatively simple point is that
the whole network breaks up in to one or more *components*, and all
the words in a component are connected to each other by paths.
Once you have succeeded in joining a word to one of these
components, you can easily join it to all the other words in that
component.

How many components are there? A theorem proved by Paul
Erdős and Alfred Rényi in 1960 implies that if on average each word
connects to enough others – more than some critical amount – then
we should expect to find one giant component containing almost all
the words, and a scattering of much smaller ones. And this is what
happens. The giant component usually misses some bits out – for

instance, if we can find an isolated word, one that has *no* immediate neighbours, then that word on its own would form one component, disconnected from everything else.

What about an obscure word like SCRY (meaning to crystal-gaze)? Is that isolated? No, SCRY connects to SPRY, then to SPAY, then SPAR, SPAN, ... and it has clearly 'escaped', with lots of potential links, so we expect it to join up to the giant component, even though my picture doesn't show how. In fact, SPAY–SPAT–SPOT–SHOT will do. This is *why* there is probably only one giant component. Since it's so big, anything that is linked by a path to a reasonable number of words has more and more potential links, and at some point the path will run into the giant component.

Ted Johnson analysed the network of four-letter words, with one slight change to the definition of a link: you are also allowed to reverse the word. This probably does not change the components significantly, if at all, because relatively few words are meaningful when reversed.

He obtained his list of four-letter words from an online dictionary, resulting in a total of 4,776. Using mathematical methods (the Graph module for the computer scripting language Perl) he found that some words are isolated (like HYMN, according to the Scrabble dictionary) or form isolated pairs. Another small component contains just eight words. That left 4,439 words: one giant component with 4,436 words, and a small one with three – TYUM, TIUM, TUUM. These are not in the Scrabble Dictionary, but *tuum* is a literary word for 'yours', from the Latin, as in 'meum and tuum' – mine and yours. I'm inclined to rule out the other two and count tuum as a single isolated word. His results can be found at: users.rcn.com/ted.johnson/fourletter.htm

If you play around with the network, you start to notice some regular structural features. The group of words BAND, BEND, BIND, BOND is an example: they are all connected to each other. That's because all the changes involve the same letter position, the second from the left. Biologists working on genetic networks call common small sub-networks *motifs*. There are five-word motifs like this, too: MARE, MERE, MIRE, MORE, MURE is one example.

A more significant motif in the word network is a series of three words like SHOT–SOOT–SORT with two vowels in the middle word. Vowels are crucial. Most single-letter changes to words change a consonant to another consonant or a vowel to another vowel. If all changes were like that, the vowel positions could never move. So changing SHIP, with a vowel in position 3, to DOCK, with a vowel in position 2, would be impossible. But sometimes consonants can change to vowels, or vowels to consonants. A sequence like SHOT–SOOT–SORT in effect moves the location of the vowel, by introducing another one and then losing the first.

In going from ORDER to CHAOS, the biggest problem is moving the vowel positions around. That's where EIDER and SOOTS come in, in fact. But notice that although both the start and end words have a vowel in position 4, some of the intermediate words don't. Sometimes you have to take a detour to get where you want to go.

Provided we take a relaxed view of what constitutes a vowel, every English word contains one. The standard vowels are AEIOU, of course. But the Y in SPRY acts like a vowel, for instance, and Y is often included in the list of vowels. The same goes for the W in the Welsh word CWM (which comes into the 4-letter network if we use the plural CWMS). If we define vowels that way, or exclude words without vowels, then the Ship–Dock theorem holds. This states that when going from SHIP to DOCK, some intermediate word must contain two vowels.

Why? At each stage the number of vowels can change by at most 1, and if it does not change, then the vowel stays in the same position. If the vowel count was always 1, then the vowel in SHIP would have to stay in the third position – but the vowel in DOCK is in the second position. So the vowel count must change. Look at the first word where it changes. It starts at 1 and changes by 1, leading to either 2 or 0. But 0 is ruled out by our convention about what constitutes a vowel or a permissible word, so it must be 2. The same theorem holds for words of any length. If the initial word has a vowel where the final one has a consonant, or vice versa, then somewhere we must hit two or more vowels. Why more? Because of examples like ARISE–AROSE, where the start and end words have more than two vowels.

• •

The Hairy Ball Theorem

Yes, a hairy ball can be combed so that it is smooth at every point
except one. The idea is to move the two tufts in the picture together
until they coincide.

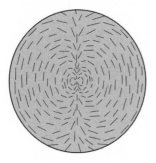

Extend the loops
smoothly over the back.

• •

Cups and Downs

Cups Puzzle 1

This one is impossible, and again the proof is parity. We start with an
even number of upright cups (zero) and end with an odd number
(11). But we are inverting an even number of cups each time, and
that implies that the parity cannot change.

Cups Puzzle 2

This time there is a solution, and the shortest takes four moves.

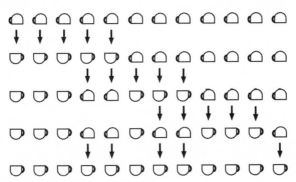

Inverting 12 cups, 5 at a time.

There's a general version using n cups, initially all upside down, where each move inverts precisely m cups. Parity rules out any solutions when n is odd and m is even. In all other cases, solutions exist. Man-Keung Siu and I proved that the shortest solution depends in a surprisingly complicated way on m and n, and there are six different cases. For the record, the answers are:

n even, m even, $2m \leq n$: $\lceil n/m \rceil$

n even, m even, $2m > n$:   1 if $m = n$, 3 if $m < n$

n even, m odd, $2m \leq n$: $2\lceil n/2m \rceil$

n even, m odd, $2m > n$: $2\lceil n/2(n-m) \rceil$

n odd, m odd, $2m \leq n$: $2\lceil (n-m)/2m \rceil + 1$

n odd, m odd, $2m > n$:   1 if $m = n$, 3 if $m < n$

Here $\lceil x \rceil$ is the *ceiling function*: the smallest integer greater than or equal to x.

• •

Secret Codes That Can Be Made Public

The general procedure for the RSA cryptosystem goes like this:

- Choose, once and for all, two prime numbers p and q. They should be really large, say 100 or even 200 digits each. Work out their product pq.
- Choose an integer e (for encode) between 1 and $(p-1)(q-1)$ which is not a multiple of p or q.
- Now Alice, who is sending the message N to Bob, does this:
- Encode the message N as N^e (mod pq).
- Transmit the message.

At this point, even Alice does not know how to decode a message. She knows what she sent, of course. Thanks to Euler, and some preliminary calculations when the code was set up, Bob knows a crucial fact that Alice doesn't:

- There is a unique integer d (for decode) in the same range, for which

$$de \equiv 1(\mathrm{mod}(p - 1)(q - 1))$$

and Bob knows what d is. Now he can decode Alice's message N^e (mod pq) by raising it to the power d:

- Form $(N^e)^d$ (mod pq)

Euler's theorem tells us that

$$(N^e)^d \equiv N^{ed} \equiv N(\mathrm{mod}\ pq)$$

so Bob has recovered the message N.

As a practical matter, it is relatively straightforward to choose p and q, work out pq, and then let Alice know what pq is – and what the encoding power e is. However, if everyone now forgot what p and q were, it would be impossible to work them out again! So, with the big primes actually employed, Alice can't deduce them from their product. Neither can anyone else who is not privy to the secret information that Bob knows.

● ●

Calendar Magic

x	$x+1$	$x+2$
$x+7$	$x+8$	$x+9$
$x+14$	$x+15$	$x+16$

The numbers always have this pattern.

If the smallest number is x, then the numbers in the 3×3 square are x, $x+1$, $x+2$, $x+7$, $x+8$, $x+9$, $x+14$, $x+15$, $x+16$. These add up to $9x + 72 = 9(x + 8)$. The volunteer tells Whodunni what x is. So all Whodunni has to do is add 8 and then multiply by 9. A quick way to multiply a number by 9 is to put 0 on the end and then subtract the number.

When the chosen number is 11, Whodunni adds 8 to get 19, and then works out $190 - 19 = 171$.

● ●

The Rule of Eleven

The largest such number is 9,876,524,130. The smallest is 1,024,375,869 (remember, don't start with 0).

How do we find these? Bearing the test in mind, we have to split the digits 0–9 into two distinct sets of five, so that the sums of these two sets differ by a multiple of 11. In fact, we can prove that the difference must be 11 or -11, like this. Let the sums concerned be a and b. Then $a - b$ is some multiple of 11. But $a + b$ is the sum of all digits 0–9, which is 45. Now, $a - b = (a + b) - 2b = 45 - 2b$. Since 45 is odd, and $2b$ is even, $a - b$ has to be odd. So it might be any of the numbers 11, 33, 55, ... or their negatives -11, -33, -55, ... However, both a and b lie between $0 + 1 + 2 + 3 + 4 = 10$ and $9 + 8 + 7 + 6 + 5 = 35$. So their difference lies between -25 and 25. That cuts the possibilities down to -11 and 11.

Now we can solve the equations $a - b = 11$, $a + b = 45$ (or $a - b = -11$, $a + b = 45$) for a and b. The result is that $a = 28$, $b = 17$, or $a = 17$, $b = 28$. So it remains to find all possible ways to write 17 as a sum of five distinct digits. We can make a systematic search, bearing in mind that the digits concerned can't be very big. For instance, $2 + 3 + 4 + 5 + 6 = 20$ is already too big, so the smallest digit must be 0 or 1, and so on. The upshot is that one set of five digits must be one of:

> 01259, 01268, 01349, 01358, 01367, 01457,
> 02348, 02357, 02456, 12347, 12356

The other set of five must be whatever these miss out, namely:

> 34678, 34579, 25678, 24679, 24589, 23689
> 15679, 14689, 13789, 05689, 04789

respectively.

To get the largest multiple of 11 using all ten digits, we must interleave the two sets, keeping all digits as big as possible starting from the left-hand end. We can make a good start with **987**6**5** using the pairs 34**579** and 01**268** (and only those), where I've used boldface and underlines to show which digit comes from which set. Continuing using the largest available possibilities (the so-called *greedy algorithm*) we get **9**8**7**6**5**2**4**13**0**.

The smallest number is slightly harder to find. We can't start with

0, so 1 is the next best bet. This should be followed by 0, if possible, then 2, 3, and so on. If we try to start 1̲0234 we're stuck, because the only set listed that contains 1, 2 and 4 is 12347, but this also has the 3, which ought to be in the other set. In fact, starting 1̲023 won't work because anything containing 12 also contains 0 or 3, but those have to be in the other set. So the next smallest possibility is to start 1̲024, and the smallest of all would start 1̲0243. That forces one set to be 12356 and the other 04789. Interleaving these digits in order we get 1024375869 as the smallest possibility.

The smallest multiple of 11 where the difference $a−b$ is not zero is 209. If you try the first few multiples of 11, such as 11, 22, 33, and so on, then $a−b$ is obviously 0 up to at least 99, since $a = b$. The next multiple, 110, also has $a = b$, and so do 121, 132, 143, 154, 165, 176, 187, 198. But 209 has $a = 11$, $b = 0$, so this is the smallest case.

* *

Digital Multiplication

2	1	9		2	7	3		3	2	7
4	3	8		5	4	6		6	5	4
6	5	7		8	1	9		9	8	1

* *

Common Knowledge

With three monks, all bearing blobs, the reasoning is as follows.

Aelfred thinks: If *I* don't have a blob, then Benedict sees a blob on Cyril but nothing on me. Then he will ask himself whether *he* has a blob. And what he will think is: 'If I, Benedict, do not have a blob, then Cyril sees that neither Aelfred nor I has a blob, so he will quickly deduce that *he* has a blob. But Cyril, an excellent logician, remains unembarrassed. Therefore I must have a blob.'

Now Aelfred reasons: 'Since Benedict is also an excellent logician, and has had plenty of time to work this out but remains unembarrassed, then I, Aelfred, must have a blob.' At this point, Aelfred turns crimson – as do Benedict and Cyril, who have followed similar lines of reasoning.

Now suppose, say, that there are only two monks, of whom only

Benedict has a blob. When the Abbot makes his announcement, Benedict sees that Aelfred has no blob, deduces that he must have one, and at the first ring of the bell, puts up his hand. Aelfred does not, because at that stage he's not yet sure of his own blob status.

Next, think of three monks; suppose that Benedict and Cyril have blobs, but Aelfred does not.

Benedict sees just one blob, on Cyril's head. He reasons that if he, Benedict, does not have a blob, then Cyril sees no blobs at all. So Cyril will raise his hand after the first ring. But Cyril doesn't (we'll shortly see why), so Benedict now knows that he must have a blob. So he raises his hand after ring 2.

Cyril is in exactly the same position as Benedict, since he also sees just one blob, on Benedict. So he does not raise his hand after the first ring, but does at the second.

Aelfred is in a rather different position. He sees two blobs, one on Benedict and one on Cyril. He wonders whether he, Aelfred, also has a blob. If so, then all three of them have blobs, and he knows from the previous version of the puzzle that they will all wait for ring 3 and then raise their hands. So he does not, and should not, raise his hand after ring 1 or 2. Then the other two do raise their hands: at that point he knows that he does not have a blob.

Again, a complete proof can be given by induction, thinking about the general case of n monks, m of whom have blobs. I'll spare you the details.

• •

Pickled Onion Puzzle

There were 31 pickled onions.

Suppose there are a pickled onions to start with, b after the first traveller has eaten, c after the second traveller has eaten, and d after the third traveller has eaten. Then

$$b = 2(a-1)/3, \qquad c = 2(b-2)/3, \qquad d = 2(c-3)/3$$

which we rewrite as

$$a = 3b/2 + 1, \qquad b = 3c/2 + 2, \qquad c = 3d/2 + 3$$

We are told that $d = 6$. Working backwards, $c = 12$, $b = 20$, $a = 31$.

• •

Guess the Card

At each stage, when Whodunni picks up the cards he makes sure that the selected pile is sandwiched between the other two. As a result, the chosen card works its way to the middle of the stack of cards. So on the final deal it is the middle card in the chosen pile.

• •

And Now with a Complete Pack

The first part of the trick is equivalent to asking which column the chosen card is in on the second deal. Knowing the column, and now being told the row, Whodunni can easily spot the card.

This trick is a bit transparent, but it can be dressed up to make it less obvious. It may be better performed with 30 cards, first dealing 6 rows of 5, then 5 rows of 6. It works with ab cards dealt in a rows of b, then b rows of a, for any whole numbers a and b.

• •

Halloween = Christmas

Because 31 Oct = 25 Dec. That is, 31 in base-8 numerals (octal) = 25 in base-10 numerals (decimal). In base 8 notation, 31 means $3 \times 8 + 1$, and this is 25.

• •

Rectangling the Square

I know at least two distinct solutions, not counting rotations or reflections as distinct. The first was found by M. den Hertog, the second by Bertie Smith. In the first the rectangles have sides 1×6, 2×10, 3×9, 4×7 and 5×8. In the second, the sides are 1×9, 2×8, 3×6, 4×7 and 5×10.

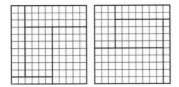

Solutions found by den Hertog and Smith.

• •

X Marks the Spot

From Abandon Hope Point: 113. From Buccaneer Bay: 99. From Cutlass Hill: 85.

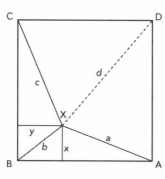

Distances for Redbeard's map.

The figure shows the three distances required: a, b, c. Of these, we know $b = 99$. Let $s = 140$ be the side of the square. Consider two further lengths x and y as shown. Then Pythagoras's theorem tells us that

$$a^2 = x^2 + (s - y)^2 = x^2 + s^2 - 2sy + y^2$$
$$c^2 = y^2 + (s - x)^2 = y^2 + s^2 - 2sx + x^2$$
$$b^2 = x^2 + y^2$$

The first step is to get rid of x and y. Subtract the third equation from the first and second, to obtain

$$2sy = s^2 + b^2 - a^2$$
$$2sx = s^2 + b^2 - c^2$$

Therefore

$$(s^2 + b^2 - a^2)^2 + (s^2 + b^2 - c^2)^2 = 4s^2(x^2 + y^2) = 4s^2 b^2$$

This is the fundamental relation between a, b, c and s.

Substitute known values $s = 140$ and $b = 99$, to get

$$(29{,}401 - a^2)^2 + (29{,}401 - c^2)^2 = 27{,}720^2$$
$$= (2^3 \times 3^2 \times 5 \times 7 \times 11)^2$$

The hint now tells us that $29{,}401 - a^2$ and $29{,}401 - c^2$ are both multiples of 7. (The corresponding statement is false for the factors 2 and 5, but true for 3 and 11.) Considering the factor 7 (a similar trick works for 3 and 11) we observe that

$$29{,}401 = 4{,}200 \times 7 + 1$$

so $1 - a^2$ and $1 - c^2$ are both multiples of 7. That is, a^2 and c^2 are of the form $7k + 1$ or $7k - 1$ for suitable integers k.

Now it is a matter of trying the possible values for a, and seeing first whether

$$23{,}800^2 - (29{,}401 - a^2)^2$$

is a perfect square, and, if so, whether the corresponding c is a whole number. The business about multiples of 7 shortens the work because the only values for a that we need check are

$$1, \quad 6, \quad 8, \quad 13, \quad 15, \quad 20, \quad 22$$

and so on. We can stop when c becomes less than a, because then we'll be trying the same calculations but with a and c interchanged.

For the $7k + 1$ case, we find $a = 85$, $c = 113$ when $k = 12$; there is also the solution $a = 113$, $c = 85$, interchanging a and c, when $k = 16$. There is no solution for the $7k - 1$ case.

Since the instructions on the back of the map say that the nearest marker is C, we want $c < a$, so $a = 113$ and $c = 85$.

That's one way to get the answer, but the mathematical story goes further.

This puzzle is a special case of the four distance problem: does there exist a square whose side is a whole number, and a point whose distances from the four corners of the square are all whole numbers? No one knows the answer. For a long time no one even knew whether three of those distances could be whole numbers.

We've already derived a relation between s, a, b and c:

$$(s^2 + b^2 - a^2)^2 + (s^2 + b^2 - c^2)^2 = (2bs)^2$$

The fourth distance d (shown dotted on my diagram) must satisfy

$$a^2 + c^2 = b^2 + d^2$$

J. A. H. Hunter discovered a formula giving some (but not all) solutions of the first equation:

$$a = m^2 - 2mn + 2n^2$$
$$b = m^2 + 2n^2$$
$$c = m^2 + 2mn + 2n^2$$
$$s^2 = 2m^2(m^2 + 4n^2)$$

and observed that s is a whole number provided that we take

$$m = 2(u^2 + 2uv - v^2)$$
$$n = u^2 - 2uv - v^2$$

for whole numbers u and v.

The choice $u = 2$, $v = 1$ leads to $s = 280$, $a = 170$, $b = 198$, $c = 226$, and we can remove the factor 2 to get $s = 140$, $a = 85$, $b = 99$, $c = 113$. The fourth side here is $d = \sqrt{10,193}$ which is not a whole number, indeed not rational. In fact, it is known that in Hunter's formula the fourth distance d can never be rational, so this formula alone won't solve the four distance problem. However, there are solutions of the three distance problem that don't arise from Hunter's formula.

This tantalising problem has deep connections with Kummer surfaces in algebraic geometry: see Richard K. Guy, *Unsolved Problems in Number Theory*.

• •

Whatever's the Antimatter?

Dirac's electron equation looks like this:

$$\left\{ p_0 + \frac{e}{c}A_0 - \rho_1\left(\boldsymbol{\sigma}, \mathbf{p} + \frac{e}{c}\mathbf{A}\right) - \rho_3 mc \right\}\psi = 0$$

where

ψ is the quantum wave-function of the electron
A_0 is the scalar potential of the electromagnetic field
$\mathbf{A}$ is the vector potential of the electromagnetic field
$\mathbf{p} = (\rho_0, \rho_1, \rho_2, \rho_3)$ is the momentum vector of the electron
$\boldsymbol{\sigma} = (\sigma_1, \sigma_2, \sigma_3)$ is a list of three 4×4 spin matrices
$\mathbf{r} = (r_1, r_2, r_3)$ is a list of three 4×4 matrices which anticommute
        with $\sigma_1, \sigma_2, \sigma_3$

e is the charge on the electron

m is the mass of the electron

c is the speed of light

Got that? I include it only to make the point that the equation is far from obvious, and because it seemed a cheat to leave it out. Everyone writes out $E = mc^2$, even Stephen Hawking,[*] and it's wrong to discriminate against formulas on grounds of complexity. Dirac spends nearly four pages of his book *The Principles of Quantum Mechanics* explaining how he derived it, and most of the previous 250 pages setting up the ideas that go into it.

Leaning Tower of Pizza

With five boxes, the maximal overhang is 1.30455. With six boxes, it is 1.4367. The stacks look like this:

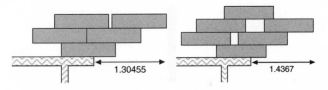

Maximal overhangs with five and six boxes.

The paper by Paterson and Zwick is 'Overhang', *American Mathematical Monthly*, vol. 116, January 2009, pp. 19–44.

PieThagoras's World-Famous Mince πs

The areas of the three pies are (mini) 9π, (midi) 16π, and (maxi) 25π, so the maxi-pie has the same area as the other two put together. If the maxi-pie is split in half, Alvin and Brenda can each have a piece $(25\pi/2)$. That leaves the other two pies to share between Casimir and Desdemona. This can be done by slicing $7\pi/2$ off the midi-pie and

---

[*] In *A Brief History of Time*, where he mentions editorial advice that every formula halves a book's sales. So he could have sold *twice as many*. Ye gods.

giving that piece plus the mini-pie to Casimir ($9\pi + 7\pi/2 = 25\pi/2$). Desdemona gets the larger piece of the midi-pie ($16\pi - 7\pi/2 = 25\pi/2$).

There are lots of ways to slice the midi-pie. The traditional one is to place the mini-pie over the middle of the midi-pie, trace round half of its circumference, and then join the ends to the edge of the midi-pie. But you could slice off any shape with area $7\pi/2$. And the cut that divides the maxi-pie in half need not be a diameter – it could be curved.

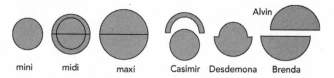

mini   midi   maxi   Casimir Desdemona Brenda

Cut the three pies as on the left, and share as on the right.

• •

Diamond Frame

There are ten basically different answers, where, for example, swapping the ace and seven in the right-hand side doesn't count as different since it's a simple transformation that obviously keeps the sums the same. There are two with a sum of 18, four with 19, two with 20, and two with 22. Here's one of them.

One of the ten answers: sums are all 18.

• •

Pour Relations

You can solve this by trial and error, or by listing all the possible states and moves and finding a path from the initial state to the desired final one. Here is one solution, which takes nine moves (two are combined in the second picture). There's a shorter solution, which I'll derive in a moment.

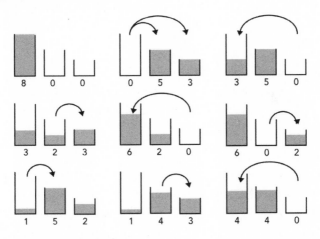

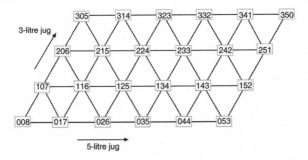

One way to divide the water.

However, there is a more systematic method, which seems to have first been published by M. C. K. Tweedie in 1939. It uses a grid of equilateral triangles, which engineers call isometric paper and mathematicians call trilinear coordinates.

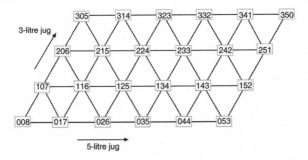

Representing the possible states of the jugs.

Here the triples of numbers indicate how much each jar holds, in the order (3-litre jug, 5-litre jug, 8-litre jug). So for example 251 means that the 3-litre jug holds 2 litres, the 5-litre jug holds 5 litres, and the 8-litre jug holds 1 litre. If you look at the first number, then the lowest line in my diagram always starts 0, the line above starts 1, and so on. Similarly, the second number reads 0, 1, 2, 3, ... from left to right in each row. So the two arrows in the diagram are 'coordinate axes' for the amount in the 3-litre jug and the amount in the 5-litre jug.

What of the 8-litre jug? Because the total amount of water is always 8 litres, the third number is always determined by the first two. Just add them and subtract from 8. But there is a nice pattern here. Thanks to the geometry of isometric paper, the third number is constant along lines that slope up and to the left – that is, the third system of lines in the picture. For example, look at the line through 035, 125, 215, 305.

If we represent the possible 'states' of the jugs – how much each contains – in this way, then the allowable moves from any state to any other take on a simple geometric form, as I'll now explain.

First, observe that the states in which some jug is either full or empty – which are the states allowed by the moves – are precisely those on the boundary of the picture.

The allowable moves, starting from some state (necessarily on the boundary) amount to moving along a line until you next hit the boundary. If you start at a corner, and move along the boundary (say from 008 to 053), then you can't stop part way along, but have to go all the way to the next corner.

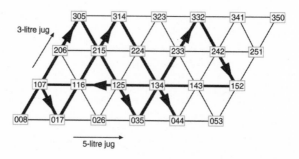

One solution.

So we can solve the puzzle by starting from 008 (lower left corner) and bouncing around the parallelogram like a billiard ball. The arrowed path shows what happens: we visit the states

008, 305, 035, 332, 152, 107, 017, 314, 044

and notice that this is the final state we want, so we stop.

This is precisely the solution given above. But there's *another* one:

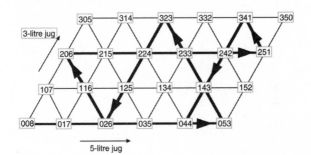

An alternative.

Now the sequence is

008, 053, 323, 026, 206, 251, 341, 044

which uses seven moves instead of nine. You may have found that one instead.

●●

The Sacred Principle of Mat

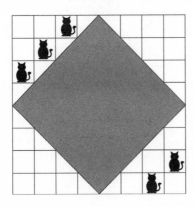

How to ensure the greatest mat.

Remember, the avatars 'watch over' every other cushion. Otherwise you could make the mat bigger.

• •

Target Practice

Tuck hit the outer ring (light grey), while Robin hit the three inner rings (dark grey).

The rings Robin and Tuck hit.

The areas of successive circles are πr^2, where $r = 1, 2, 3, 4, 5$, respectively:

$$\pi, \quad 4\pi, \quad 9\pi, \quad 16\pi, \quad 25\pi$$

The areas of the rings are the differences between these:

$$\pi, \quad 3\pi, \quad 5\pi, \quad 7\pi, \quad 9\pi$$

These are π multiplied by consecutive odd integers. Robin's integers are less than or equal to Tuck's, since Robin's arrows are closer to the centre. The two sets of odd integers must have the same sum. The only possibility is $1 + 3 + 5 = 9$.

Bonus point: Six rings. The sixth ring has area 11π, so $1 + 3 + 7 = 11$ is a second solution.

Further bonus point: Eight rings. Robin's odd numbers must be consecutive, and so must Tuck's. The next two rings have areas 13π, 15π, and $3 + 5 + 7 = 15$ is a second solution with consecutive odd numbers.

• •

Just a Phase I'm Going Through

AC is exactly $\frac{1}{2}$AB. The times at which such crescents appear are $\frac{1}{6}$ and $\frac{5}{6}$ through the lunar cycle, starting from the new moon. (*Not* $\frac{1}{4}$ *and* $\frac{3}{4}$!)

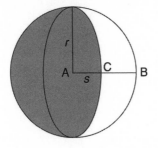

When the area of the crescent is one-quarter of the area of the disc.

The inner edge of the crescent is half of an ellipse (*Cabinet*, page 283). The full ellipse is shown. The white crescent has $\frac{1}{4}$ the area of the circle when the ellipse has $\frac{1}{2}$ the area of the circle. Let AB $= r$, AC $= s$. The area of the circle is πr^2. The area of an ellipse is πab where a and b are the 'semiaxes' – half of the widths in the widest and narrowest directions. Here $a = r$ and $b = s$. So we want $\pi rs = \frac{1}{2}\pi r^2$, so $s = \frac{1}{2}r$.

For the timing of these crescents, view the Moon from 'above'. The centre of the Earth is at E; its orbit is shown as the larger circle, but not to scale. The light from the Sun S illuminates half the Moon, leaving the other half dark (here shown in grey). New moon occurs when the centre of the Moon is at point O.

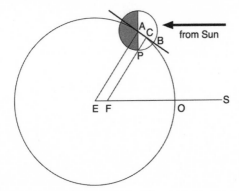

Geometry of lunar orbit.

The points A, C, B correspond to those in the question, and we want to choose angle SEA to make C the midpoint of AB, where P is the edge of the dark area of the Moon and FPC is parallel to EA (the assumption of parallel projection). Then BP = AP (since triangle APB is isosceles), but AP = AB since both are radii of the Moon. Therefore triangle APB is actually equilateral, so angle PAB = 60°. Therefore angle PAE = 30° and angle SEA = 60°, one-sixth of a full circle. The Moon is thus 1/6 of a cycle round from the new moon.

There is a corresponding position at 5/6 of a cycle, obtained by reflecting the diagram in the line ES.

How Dudeney Cooked Loyd

- The children's puzzle seems impossible on grounds of parity (odd/even). All the numbers are odd, and six odd numbers must add to an even number, not 21. Gardner's answer was to cook his own puzzle, turn the page upside down, and circle the three 6's and the three 1's that then appear. But a reader, Howard Wilkerson, circled each of the 3's, one of the 1's, and then drew a big circle round the other two 1's (giving 11). I think this is a more elegant quibble-cook.

- Loyd's construction leads to a rectangle which, while almost square, has sides of differing lengths. If the mitre is made from a square of side 1, giving it an area of $\frac{3}{4}$, then Loyd's 'square' measures 6/7 horizontally by 7/8 vertically.

- Dudeney's five-piece solution, which is exact if the lengths are chosen correctly, looks like this:

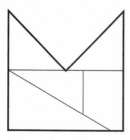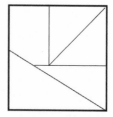

Dudeney's solution.

No one knows a four-piece dissection from the mitre to the square, and there probably isn't one, but its existence has not been ruled out.

• •

Cooking with Water

What I *really* should have said, to be super-cautious, was 'you are not allowed to pass the *connections* through a house or a utility company building'. As David Uphill pointed out, there is a way to solve the problem if the words are interpreted literally, even with pipes forbidden to pass through houses. I've modified his suggestion slightly to fit my question more closely. It uses two big water tanks to pass the water connections through two of the houses. No connecting pipe even *enters* a house, let alone passes through it.

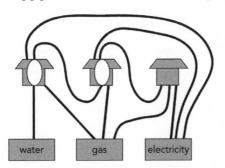

The utilities puzzle quibble-cooked.

Hmm... If you feel that a tank used like this is just a large pipe by another name, which was my first reaction, then this layout doesn't fit the conditions. That's why I consider it a quibble-cook. But it's an ingenious one and deserves to be better known.

• •

Calculator Curiosity 2

When you multiply 0588235294117647 by 2, 3, 4, 5, ..., 16, the same sequence of digits arises, in the same cyclic order. That is, you have to start at a different place, and when you get to the end, you continue from the beginning. Specifically,

$$0588235294117647 \times 2 = 1176470588235294$$
$$0588235294117647 \times 3 = 1764705882352941$$
$$0588235294117647 \times 4 = 2352941176470588$$
$$0588235294117647 \times 5 = 2941176470588235$$
$$0588235294117647 \times 6 = 3529411764705882$$
$$0588235294117647 \times 7 = 4117647058823529$$
$$0588235294117647 \times 8 = 4705882352941176$$
$$0588235294117647 \times 9 = 5294117647058823$$
$$0588235294117647 \times 10 = 5882352941176470$$
$$0588235294117647 \times 11 = 6470588235294117$$
$$0588235294117647 \times 12 = 7058823529411764$$
$$0588235294117647 \times 13 = 7647058823529411$$
$$0588235294117647 \times 14 = 8235294117647058$$
$$0588235294117647 \times 15 = 8823529411764705$$
$$0588235294117647 \times 16 = 9411764705882352$$

For my second question:

$$0588235294117647 \times 17 = 9999999999999999$$

The source of this remarkable number is the decimal expansion of the fraction 1/17, which is

0.0588235294117647 0588235294117647
0588235294117647 . . .

repeating indefinitely.

● ●

Which is Bigger?

By direct calculation, $e^\pi = 23.1407$ whereas $\pi^e = 22.4592$. So $e^\pi > \pi^e$.

Actually, there's a more general result: $e^x \geq x^e$ for any number $x \geq 0$, and equality holds if and only if $x = e$. So, not only is $e^\pi > \pi^e$, but $e^2 > 2^e$, $e^3 > 3^e$, $e^4 > 4^e$, $e^{\sqrt{2}} > (\sqrt{2})^e$, and indeed $e^{999} > 999^e$. The simplest proof uses calculus, and here it is for anyone who wants to see the details. It also helps to explain why e^π and π^e are so close together.

Let $y = x^e e^{-x}$, where $x \geq 0$. We find the stationary points (maxima, minima, and the like) by setting $dy/dx = 0$. Now

$$\frac{dy}{dx} = (ex^{e-1} - x^e)e^{-x}$$

which is zero at $x = 0$ and $x = e$, and nowhere else. The value of y at $x = 0$ is $y = 0$, which is clearly a minimum; the value at $x = e$ is $y = 1$. This is in fact a maximum. To see why, calculate the second derivative

$$\frac{d^2y}{dx^2} = [e(e-1)x^{e-2} - 2ex^{e-1} + x^e]e^{-x}$$

at $x = e$. This works out as -1, which is negative, so $x = e$ is a maximum, and the maximum value of y is 1.

Therefore $x^e e^{-x} \leq 1$ for all $x \geq 0$, with equality only at the unique maximum $x = e$. Multiply both sides by e^x, to get

$$x^e \leq e^x$$

for all $x \geq 0$, with equality only when $x = e$. Done!

The graph of the function $y = x^e e^{-x}$ has a single peak at $x = e$, and tails off towards zero as x increases to infinity.

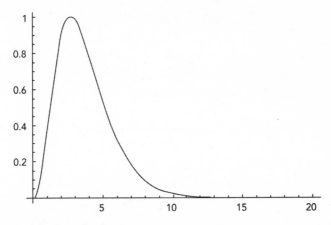

Graph of $y = x^e e^{-x}$.

This helps to explain why e^π and π^e are so close together for it not to be immediately obvious which is bigger. The graph also shows that if a number x is reasonably close to e, then $x^e e^{-x}$ is close to 1, so x^e is close to e^x. For example, if x lies between 1.8 and 3.9, then x^e is at least $0.8e^x$. In particular, this holds for $x = \pi$.

• •

Colorado Smith and the Solar Temple

The division shown solves the puzzle. So does its reflection in the diagonal.

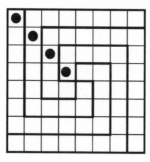

Four regions of the same shape, each containing a sun-disc.

• •

Why Can't I Add Fractions Like I Multiply Them?

The short response is that we can't add fractions that way because we don't get the right answer! Since $\frac{3}{7}$ is nearly $\frac{1}{2}$, and so is $\frac{2}{5}$, then when we add them the result must be at least $\frac{1}{2}$. But $\frac{5}{12}$ is less than $\frac{1}{2}$ because half of 12 is 6. The error is even more glaring when we try it on $\frac{1}{2} + \frac{1}{2}$, because

$$\frac{1}{2} + \frac{1}{2} = \frac{1+1}{2+2} = \frac{2}{4}$$

makes no sense: since $\frac{2}{4} = \frac{1}{2}$ it tells us that $\frac{1}{2} + \frac{1}{2} = \frac{1}{2}$.

All very well, but why does the multiplication rule work, and what should we use for addition instead?

The easy way to see why the rules differ – and what they should be – is to use pictures. Here's a picture for $\frac{2}{5} \times \frac{3}{7}$.

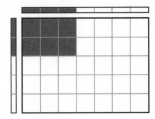

Multiplying fractions.

The vertical bar shows a line of five equal pieces, with two of them shaded grey. That represents $\frac{2}{5}$: two parts out of five. Similarly the horizontal bar represents $\frac{3}{7}$. The rectangles represent multiplication, because the area of a rectangle is what you get when you multiply the two sides. The big rectangle contains $5 \times 7 = 35$ squares. The shaded one contains $2 \times 3 = 6$ squares. So the shaded rectangle is $\frac{6}{35}$ of the big one.

When it comes to addition, the corresponding picture looks like this:

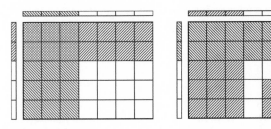

Adding fractions.

We get $\frac{2}{5}$ of the big rectangle by taking the top two rows out of five, and $\frac{3}{7}$ by taking the left-hand three columns out of seven. These regions are shown in the left-hand picture, with different shading, and they *overlap*. To count how many squares there are altogether, we either count the overlapping squares twice, or make an extra copy as in the right-hand picture. Either way, we get 29 squares out of 35, so the sum must be

$$\frac{2}{5} + \frac{3}{7} = \frac{29}{35}$$

To see how the 29 relates to the original numbers, just count the squares in the top two rows, 2×7, and add those in the left-hand three columns, 3×5. Then $2 \times 7 + 3 \times 5 = 29$. So the addition rule is

$$\frac{2}{5} + \frac{3}{7} = \frac{2 \times 7 + 3 \times 5}{5 \times 7}$$

This is where the usual recipe 'put both fractions over the same denominator' comes from.

• •

Pooling Resources

It was a bad idea. Separately, Christine would make £150 and Daphne £100, a total of £250. Together, their total income would be £240 – which is smaller.

Both pairs of ladies are making an unwarranted assumption; it happens to work in favour of the first pair and against the second pair. The assumption is that the way to combine prices of a for £b and c for £d is to add the numbers, getting $a + c$ for £$(b + d)$. This boils down to trying to add the corresponding fractions using the rule

$$\frac{a}{b} + \frac{c}{d} = \frac{a + b}{c + d}$$

and we have seen with the previous puzzles that this doesn't work. Sometimes it is an underestimate, sometimes an overestimate. It is correct when the two fractions concerned are the same.

• •

Welcome to the Rep-Tile House

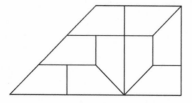

This shape is rep-9.

• •

Cooking on a Torus

A torus can be represented as a rectangle with opposite faces identified – that is, 'wrapped round' so that anything that disappears off one edge reappears at the opposite edge. A Möbius band can be drawn as a rectangle with the left- and right-hand edges identified, but with a half twist. Drawn that way, here are possible solutions. Remember, if you draw it on a Möbius band made from paper, then the lines are deemed to 'soak through' the sheet of paper.

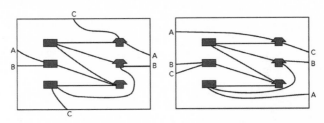

Connecting utilities on a torus... ...and a Möbius band.

• •

The Ham Sandwich Theorem

Lots of examples prove that in general you can't bisect three shapes with a single straight line. Here's one with three circles. It's easy to show that the *only* line bisecting the lower two is the one illustrated. But this doesn't bisect the third one.

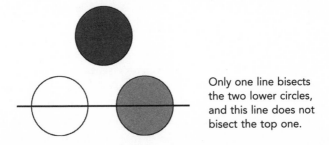

Only one line bisects the two lower circles, and this line does not bisect the top one.

In three dimensions, exactly the same idea works with four spheres. The centres of three of them lie on some plane, and provided those centres are not in a straight line, there is exactly one such plane. Now put the centre of the fourth sphere at a point that is not on the plane.

• •

Cricket on Grumpius

The Grumpians are septimists, and use base-7 arithmetic. In their system, 100 works out as

$$1 \times 7^2 + 0 \times 7 + 0 \times 1 = 49$$

So they get very excited instead of being disappointed: the batsthing with decimal 49 runs has just scored a Grumpian century!

• •

The Missing Piece

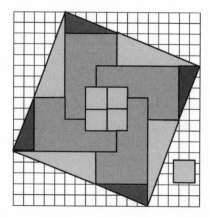

Innumeratus's solution.

Well, it does look pretty convincing ... but something must be wrong, because the area of Innumeratus's 'square' must be less than that of the original 'square'. In fact, neither shape is a perfect square. The original one bulges slightly outwards in the middle; the second bulges slightly inwards. For example, the two different sized triangles have horizontal and vertical sides in the ratios 8:3 and 5:2 respectively. If the figure were square, these ratios would be equal. But they are 2.67 and 2.5, which are different.

• •

Pieces of Five

The bosun placed one coin on the table, and then put two on top of it so that they touched at its centre. To do this, hold them in place while you fit the other two coins almost on edge, leaning together to touch at the top. Again, all coins touch, so in particular they are equidistant.

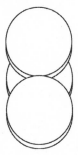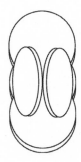

Place the first three coins as on the left, then add the other two.

• •

The Curious Incident of the Dog

The next number in the sequence is 46.

Holmes's point is: don't look at what's there, look at what's *missing*. The missing numbers are:

> 3 5 6 9 10 12 13 15 18 20 21 23 24 25 27
> 30 31 32 33 34 35 36 37 38 39 40 42 43

These are the multiples of 3, the multiples of 5, anything containing a digit 3, and anything containing a digit 5. The next number in the sequence is therefore 46 (because 45 is a multiple of 5).

• •

Mathematics Made Difficult

Lagrange's interpolation formula states that the polynomial

$$P(x) = \sum_{j=1}^{n} y_j \prod_{\substack{k=1 \\ k \neq j}}^{n} \frac{x - x_k}{x_j - x_k}$$

satisfies $P(x_j) = y_j$ for $j = 1, \ldots, n$. Remember: Linderholm's book is based on the premise that mathematics should be made as complicated as possible to enhance the prestige of the mathematician. Actually, the basic idea here is simple. In less compact notation, the formula becomes

$$P(x) = \frac{(x - x_2)(x - x_3) \cdots (x - x_n)}{(x_1 - x_2)(x_1 - x_3) \cdots (x_1 - x_n)} y_1 + \cdots$$

$$+ \frac{(x - x_1)(x - x_2) \cdots (x - x_{n-1})}{(x_n - x_1)(x_n - x_2) \cdots (x_n - x_{n-1})} y_n$$

When $x = x_j$, all terms except the jth vanish because of the factor $(x - x_j)$. The jth term doesn't have that factor, and it is an apparently complicated fraction times y_j. However, the numerator and denominator of the fraction are identical, so the fraction is 1. And 1 times y_j is y_j. Cunning!

For example, to justify the sequence 1, 2, 3, 4, 5, 19, we take $x_1 = 1$, $x_2 = 2$, $x_3 = 3$, $x_4 = 4$, $x_5 = 5$, $x_6 = 6$ and $y_1 = 1$, $y_2 = 2$, $y_3 = 3$, $y_4 = 4$, $y_5 = 5$, $y_6 = 19$. Then a calculation gives

$$P(x) = \frac{13}{120} x^5 - \frac{13}{8} x^4 + \frac{221}{24} x^3 - \frac{195}{8} x^2 + \frac{1841}{60} x - 13$$

and $P(1) = 1$, $P(2) = 2$, $P(3) = 3$, $P(4) = 4$, $P(5) = 5$, $P(6) = 19$.

Edward Waring first published the formula in 1779. Euler rediscovered it in 1783, and Lagrange discovered it again in 1795. So it is named after the third person who found it, which is fairly typical when it comes to naming mathematical ideas after people.

• •

A Four Colour Theorem

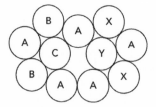

These 11 circles require four colours.

Eleven circles are needed. Here's an arrangement that requires four colours. Suppose for a contradiction that it can be 3-coloured. Colour the top middle circle with colour A, and the two adjacent ones with colours B and C. Then the next circle to the left must have colour A, the one below that colour B, and the left-hand circle of the two lowest ones must have colour A. The colours round the right-hand side of the figure must be similar: here either $X = B$ and $Y = C$, or $X = C$ and $Y = B$. Either way, the right-hand circle of the two lowest ones must have colour A. But now two adjacent circles have the same colour, namely A – contradiction.

It can be proved that with ten or fewer circles, at most three colours are needed.

• •

Serpent of Perpetual Darkness

The Earth goes round the Sun exactly once in a year, so it returns to the same point in its orbit on any given date (subject to a bit of drift because the exact period is not an integer number of days, whence our use of leap years). In particular, every 13 April it returns to a position at which the orbit of Apophis crosses that of Earth, a necessary condition for a collision.

Apophis has a period of 323 days, and the distance between Apophis and the Sun ranges from 0.7 astronomical units to 1.1 astronomical units, where an astronomical unit is the average distance between the Sun and the Earth. So sometimes it is inside Earth's orbit, sometimes outside. If Apophis and the Earth orbited in the same plane, their orbits would cross at two points. However, it's not quite that simple. The orbits are inclined at a small angle, and Apophis is light enough to be affected significantly by the gravitational pull of the other planets, causing changes to its orbit. So although the orbits don't necessarily *cross*, the orbit of Apophis – though not necessarily the asteroid itself – comes close to that of the Earth in two places, and the Earth reaches those positions on two specific dates. The one that counts, for the near future, is where the Earth is every 13 April. Whether there's a collision depends on precisely where Apophis is in its orbit on that date, and a lot of high-precision observations are needed to determine that. So the date is easy, but the year is difficult.

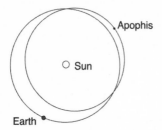

Orbit of Apophis relative to that of the Earth.

Actually, there was more to it. There always is. Calculations showed that if Apophis happens to pass through a particular region of space, about 600 metres across, during a near-miss in 2029, then it is guaranteed to return to almost the same spot, hitting the Earth in 2036. Fortunately, the latest observations indicate that the chance of such a collision is at most 1 in 45,000. See

neo.jpl.nasa.gov/news/news146.html

science.nasa.gov/headlines/y2005/13may_2004mn4.htm

Both of these refer to Apophis by its provisional designation 2004 MN$_4$.

●●●

What Are the Odds?

No, she's wrong, the probability is $\frac{2}{3}$. And it's very naughty of her to try to swindle poor Innumeratus like that.

Whichever card he picks first, the remaining cards include two of the opposite colour but only one of the same colour. So the probability of picking a card of the opposite colour is $\frac{2}{3}$. Since this holds whichever card he picks first, the probability that his two cards are different is $\frac{2}{3}$.

Here's another way to see it. There are 6 distinct pairs of cards. Of these, precisely 2 (♠♣ and ♥♦) have the same colours, and 4 have different colours. So the probability of getting one of these four is $\frac{4}{6} = \frac{2}{3}$.

●●●

The Shortest Mathematical Joke Ever

In analysis, ε is *always* taken to be a small *positive* number, to the point of cliché. So this joke is a more intellectual variant on the mechanics question that began 'An elephant whose mass can be neglected ...'

● ●

Name the Cards

The cards were the King of Spades, Queen of Spades and Queen of Hearts. The first has to be a Spade and the third a Queen, but the precise sequence is not determined.

The first two statements tell us that the cards must either be KQQ or QKQ.

The last two statements tell us that the cards must either be ♠♠♥ or ♠♥♠.

Combining these we find four possible arrangements:

K♠ Q♠ Q♥

K♠ Q♥ Q♠

Q♠ K♠ Q♥

Q♠ K♥ Q♠

The fourth of these contains the same card twice, so it is ruled out. The other three all use the same three cards, in various orders.

This puzzle was invented by Gerald Kaufman.

● ●

Nice Little Earner

Surprisingly, Smith earned more – even though £1,600 per year is greater than Smith's accumulated £500 + £1,000 over a year. To see why, tabulate their earnings for each six-month period:

	Smith	Jones
Year 1 first half	£5,000	£5,000
Year 1 second half	£5,500	£5,000
Year 2 first half	£6,000	£5,800
Year 2 second half	£6,500	£5,800
Year 3 first half	£7,000	£6,600
Year 3 second half	£7,500	£6,600

Note that Jones's £1,600 splits into two amounts of £800 for each half-year, so his half-year figures increase by £800 every year. Smith's half-year figures increase by £500 every half-year. Despite that, Smith is ahead in every period after the first, and gets ever further ahead as time passes. In fact, at the end of year n, Smith has earned a total of $10,000n + 500n(2n - 1)$ pounds, while Jones has earned a total of $10,000n + 800n(n - 1)$ pounds. So Smith $-$ Jones $= 200n^2 + 300n$, which is positive and grows with n.

• •

A Puzzle for Leonardo

Emperor Frederick II was seeking a rational number x such that x, $x - 5$ and $x + 5$ are all perfect squares. The simplest solution is

$$x = \frac{1,681}{144} = \left(\frac{41}{12}\right)^2$$

for which

$$x - 5 = \frac{961}{144} = \left(\frac{31}{12}\right)^2$$

$$x + 5 = \frac{2,401}{144} = \left(\frac{49}{12}\right)^2$$

Leonardo explained his solution in 1225 in his *Book of Squares*. In modern notation, he found a general solution

$$\left(\frac{m^2 + n^2}{2}\right)^2 - mn(m^2 - n^2) = \left(\frac{m^2 - 2mn - n^2}{2}\right)^2$$

$$\left(\frac{m^2 + n^2}{2}\right)^2 + mn(m^2 - n^2) = \left(\frac{m^2 + 2mn - n^2}{2}\right)^2$$

Here the role of x is played by the number $\frac{1}{2}(m^2 + n^2)$, and we want $mn(m^2 - n^2) = 5$. Choosing $m = 5$, $n = 4$, we get $x = 3\frac{1}{2}$ and $mn(m^2 - n^2) = 180$. This may not seem much help, but $180 = 5 \times 6^2$. Dividing x by 6 yields the answer.

• •

It's About Time

Crossnumber solution.

● ●

Do I Avoid Kangaroos?

I avoid kangaroos.

Write the conditions symbolically, as on page 275. Let

A = avoided by me
C = cat
D = detested by me
E = eats meat
H = in this house
K = kangaroos
L = loves to gaze at the moon
M = kills mice
P = prowls by night
S = suitable for pets
T = takes to me

Then with ⇒ meaning 'implies' and ¬ meaning 'not', the statements (in order) become

$$H \Rightarrow C, \quad L \Rightarrow S, \quad D \Rightarrow A, \quad E \Rightarrow P, \quad C \Rightarrow M$$
$$T \Rightarrow H, \quad K \Rightarrow \neg S, \quad M \Rightarrow E, \quad \neg T \Rightarrow D, \quad P \Rightarrow L$$

Now we appeal to the laws of logic that I mentioned on page XXX:

X ⇒ Y is the same as ¬ Y ⇒ ¬ X
If X ⇒ Y ⇒ Z, then X ⇒ Z

Using these laws, we can rewrite these conditions as

¬ A ⇒ ¬ D ⇒ T ⇒ H⇒ C ⇒ M ⇒ E ⇒ P ⇒ L ⇒ S ⇒ ¬ K
so that ¬ A ⇒ ¬ K, or equivalently, K ⇒ A.

Therefore I avoid kangaroos.

• •

The Klein Bottle

To cut a Klein bottle into two Möbius bands, slice it lengthwise, cutting through the 'handle' of the bottle and the body along the plane of mirror symmetry. A little thought shows that the two pieces are Möbius bands.

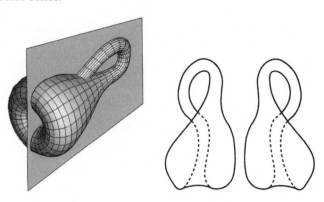

Cutting a Klein bottle into two Möbius bands.

• •

Accounting the Digits

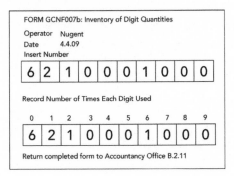

FORM GCNF007b: Inventory of Digit Quantities

Operator   Nugent
Date   4.4.09
Insert Number

6	2	1	0	0	0	1	0	0	0

Record Number of Times Each Digit Used

0	1	2	3	4	5	6	7	8	9
6	2	1	0	0	0	1	0	0	0

Return completed form to Accountancy Office B.2.11

This is the only number that works.

As Long as I Gaze on Laplacian Sunrise

If Laplace's figures are correct – which is highly debatable – the probability that the Sun will always rise is *zero*.

The probability of the Sun rising on day n is $(n-1)/n$.

So:

- The probability of the Sun rising on day 2 is $\frac{1}{2}$
- The probability of the Sun rising on day 3 is $\frac{2}{3}$
- The probability of the Sun rising on day 4 is $\frac{3}{4}$

and so on. Therefore

- The probability of the Sun rising on days 2 and 3 is $\frac{1}{2} \times \frac{2}{3} = \frac{1}{3}$
- The probability of the Sun rising on days 2, 3 and 4 is $\frac{1}{3} \times \frac{3}{4} = \frac{1}{4}$
- The probability of the Sun rising on days 2, 3, 4 and 5 is $\frac{1}{4} \times \frac{4}{5} = \frac{1}{5}$

and so on. The pattern is clear (and easy to prove): the probability of the Sun rising on all of the days 2, 3, ..., n is $1/n$. As n becomes arbitrarily large, this tends to 0.

Strictly for Calculus Buffs

For the details, see:

D. P.Dalzell, 'On 22/7', *Journal of the London Mathematical Society*, vol. 19 (1944), pp. 133–34.

Stephen K. Lucas, 'Approximations to π derived from integrals with nonnegative integrands', *American Mathematical Monthly*, vol. 116 (2009), pp. 166–72.

• •

The Statue of Pallas Athene

The statue contained 40 talents of gold.

The four fractions add up to give

$$\frac{1}{2} + \frac{1}{8} + \frac{1}{10} + \frac{1}{20} = \frac{20 + 5 + 4 + 2}{40} = \frac{31}{40}$$

so what's left is $\frac{9}{40}$. Since this requires 9 talents, the total must have been 40 talents.

• •

Calculator Curiosity 3

$6 \times 6 = 36$
$66 \times 66 = 4356$
$666 \times 666 = = 443556$
$6666 \times 6666 = 44435556$
$66666 \times 66666 = 4444355556$
$666666 \times 666666 = 444443555556$
$6666666 \times 6666666 = 44444435555556$
$66666666 \times 66666666 = 4444444355555556$

• •

Completing the Square

The stated conditions do not require you to use the integers 1–9, and in fact no solution exists if you do, because then the even numbers must go in the corners. But by using fractions, you can solve the puzzle. The figure shows the traditional solution, perhaps the simplest, but there are infinitely many others, even if you restrict the entries to positive numbers.

$4\frac{1}{2}$	8	$2\frac{1}{2}$
3	5	7
$7\frac{1}{2}$	2	$5\frac{1}{2}$

An unorthodox
magic square.

The Look and Say Sequence

The rule of formation is best stated in words. The first term is '1', which can be read as 'one one', so the next term is 11. This reads 'two ones', leading to 21. Read this as 'one two, one one' and you see where 1211 comes from, and so on.

Conway proved that if $L(n)$ is the length of the nth term in this sequence, then

$$L(n) \approx (1.30357726903\ldots)^n$$

where $1.30357726903\ldots$ is the smallest real solution of the 71st-degree polynomial equation

$$\begin{aligned}
&x^{71} - x^{69} - 2x^{68} - x^{67} + 2x^{66} + 2x^{65} - x^{63} - x^{62} - x^{61} - x^{60} \\
&+ 2x^{58} + 5x^{57} + 3x^{56} - 2x^{55} - 10x^{54} - 3x^{53} - 2x^{52} + 6x^{51} \\
&+ 6x^{50} + x^{49} + 9x^{48} - 3x^{47} - 7x^{46} - 8x^{45} - 8x^{44} + 10x^{43} \\
&+ 6x^{42} + 8x^{41} - 5x^{40} - 12x^{39} + 7x^{38} - 7x^{37} + 7x^{36} + x^{35} \\
&- 3x^{34} + 10x^{33} + x^{32} - 6x^{31} - 2x^{30} - 10x^{29} - 3x^{28} + 2x^{27} \\
&+ 9x^{26} - 3x^{25} + 14x^{24} - 8x^{23} - 7x^{21} + 9x^{20} + 3x^{19} - 4x^{18} \\
&- 10x^{17} - 7x^{16} + 12x^{15} + 7x^{14} + 2x^{13} - 12x^{12} - 4x^{11} - 2x^{10} \\
&+ 5x^9 + x^7 - 7x^6 + 7x^5 - 4x^4 + 12x^3 - 6x^2 + 3x - 6 \\
&= 0
\end{aligned}$$

I don't claim this is obvious.

The Millionth Digit

The millionth digit is 1.

> The numbers 1–9 occupy the first 9 positions.
> The numbers 10–99 occupy the next $2 \times 90 = 180$ positions.
> The numbers 100–999 occupy the next $3 \times 900 = 2,700$ positions.
> The numbers 1,000–9,999 occupy the next $4 \times 9,000 = 36,000$ positions.
> The numbers 10,000–99,999 occupy the next $5 \times 90,000 = 450,000$ positions.

At this stage, we have reached the 488,889th digit altogether. Since $1,000,000 - 488,889 = 511,111$, we want the digit in the 511,111th place, in the block that starts 100,000-100,001-100,002- and so on. Since these are grouped in sixes, we work out $511,111/6 = 85,185 \frac{1}{6}$. Therefore we are seeking the first digit of the 85,186th 6-digit block. That block must be 185,185, and its first digit is 1.

• •

Piratical Pathways

Redbeard's bank is at 19 Taxhaven Street.

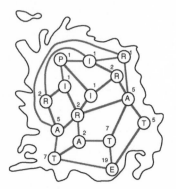

Calculating the number of paths.

The number is small enough that you can just list the possible paths, but there's a systematic method for this sort of question. The diagram shows the same map; I've removed superfluous connections that can never be used to keep thing simple, but it makes no difference to the method or the result if you leave them in.

I've written numbers beside the letters. These numbers tell us how many ways there are to reach that particular letter, and we calculate them in turn: P, the three I's, the four R's, the three A's, the three T's, and the final E.

- Start by writing 1 next to the P.
- There is exactly one path from the P to each of the I's, so we write 1 next to each I.
- Look at each R in turn, see which I's connect to it, and add up the numbers beside those letters. Here one of the R's is connected to only one I, numbered 1, so it also gets the number 1. The other three are connected to two I's, numbered 1, so they are given the number $1 + 1 = 2$.
- Next, move on to the A's. The leftmost A is connected to three R's: one numbered 1 and two numbered 2, so we give that A the number $1 + 2 + 2 = 5$. And so on.
- Continuing in this way, we eventually reach the final E. The T's that connect to it are numbered 7, 7 and 5. So we number E with $7 + 7 + 5 = 19$. And that's the number of ways to get to E.

• •

Trains That Pass in the Siding

Yes, they can pass each other – however long the trains might be.

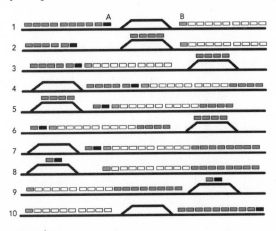

Here's how.

1 Initially, each train is on its side of the siding.
2 Train B backs away to the right. Train A runs right, past the siding, backs on to it, drops off four coaches, returns to the main line going right, and backs off to the far left.
3 Train A moves past the siding, going left, and joins the main part of train B.
4 Trains A+B move right, going along the siding, pick up the four coaches, and return to the main line on the right of the siding.
5 Then they back up on to the siding, drop off four more coaches, and return to the main line on the right of the siding.
6 The main part of the combined trains A+B moves left along the main line until it clears the siding.
7 Again trains A+B move right, going along the siding, pick up the four coaches, and return to the main line on the right of the siding.
8 Then they back up on to the siding, drop off one coach and locomotive A, and return to the main line on the right of the siding.
9 The main part of the combined trains A+B moves left along the main line until it clears the siding.
10 Finally, A+B goes right along the siding, rejoins with locomotive A and its coach. Then the trains split apart and each continues on its way.

The same method works no matter how long the trains are, provided the siding can contain at least one coach or locomotive.

• •

Squares, Lists and Digital Sums

The next such sequence is

$$99{,}980{,}001, \quad 100{,}000{,}000, \quad 100{,}020{,}001, \quad 100{,}040{,}004$$
$$100{,}060{,}009, \quad 100{,}080{,}016, \quad 100{,}100{,}025$$

which are the squares of the numbers 9,999–10,005.

A good place to look is the squares of numbers 100...00,

100...01, 100...02, 100...03, 100...04, 100...05, which have lots of zeros, while the few remaining digits give digital sums that are the squares 1, 4, 9, 16, 16, 9. To extend this list of six consecutive squares to seven, we have to look at 999...9 and 100...06. The digits of 99^2 sum to 18, not a square; those of 999^2 sum to 27, also not a square. But the digits of 9999^2 sum to 36, a square. Looking at the other end, the digits of 106^2, $1,006^2$, and $10,006^2$ sum to 13, not a square.

To rule out anything in between 15^2 and 9999^2, we just have to find a sequence of squares of numbers that differ by at most 6, whose digits sum to non-squares. For example,

$16^2 = 256$ with digit-sum 13
$19^2 = 361$ with digit-sum 10
(20^2, 21^2, and 22^2 have square digit-sums so I can't use those)
$25^2 = 625$ with digit-sum 13
$29^2 = 841$ with digit-sum 13

and so on. I'm sure there must be short cuts, and a computer can quickly check all possibilities in that range.

No one seems to know whether *eight* consecutive squares can all have square digit-sums.

• •

Match Trick

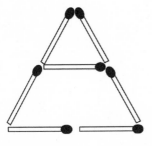

It's easy if you let the edges of the triangles overlap.

• •

Slicing the Cake

You can get at most 16 pieces. Here's one way to do it:

How to get
16 pieces with five cuts.

In general, using n cuts, the maximum number of pieces is $\frac{1}{2}n(n + 1) + 1$, the nth triangular number plus 1.

. . . .

Sliding Coins

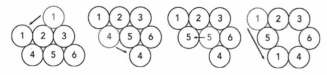

Like this.

Note that on the third move, coin 5 just slides out from between coins 2 and 4. The arrow doesn't show the direction of the move, just which coin goes where.

. . . .

Beat That!

One of the dice stopped rolling with a 6 uppermost. The other hit a rock, split in two, and the two pieces showed a 6 and a 1. So Olaf scored 13, beating the King of Sweden's feeble 12.

In mathematical circles this kind of thing is called 'expanding the state space'. That is, extending the range of possible outcomes. It is one of the reasons why mathematical models never match reality perfectly.

In gambling circles it's called 'rigging the dice'.

In political circles it's called 'politics'.

I learned this story from Ivar Ekeland's *The Broken Dice*.

• •

Euclid's Puzzle

The donkey was carrying five sacks and the mule was carrying seven.

Suppose the donkey carries x sacks and the mule y. Then the mule tells us two things:

$$y + 1 = 2(x - 1)$$
$$x + 1 = y - 1$$

The second equation tells us that $y = x + 2$. Now the first equation tells us that $x + 3 = 2x - 2$, which implies that $x = 5$. So $y = 7$.

• •

The Infinite Monkey Theorem

Each character has a probability of 1/36 of being selected on any given throw, so on average it takes 36 throws to get any given character. To get DEAR SIR, with 8 characters including the space, you need

$$36 \times 36 \times 36 \times 36 \times 36 \times 36 \times 36 \times 36 = 36^8$$
$$= 2,821,109,907,456$$

throws. The complete works of Shakespeare would need $36^{5000000}$ throws, which is roughly $10^{2385606}$. If the monkey typed 10 characters per second, which is faster than a really good typist, it would take roughly 3×10^{2385597} years to complete the task.

Dan Oliver ran a computer program in 2004, and after a simulated time of 42 octillion years, the digital monkey typed
VALENTINE. Cease toIdor:eFLP0FRjWK78aXzVOwm)-';8.t

The first 19 letters appear in *The Two Gentlemen of Verona*. Similar results are reported at:

en.wikipedia.org/wiki/Infinite_monkey_theorem

• •

Snakes and Adders

If a square board with no missing corner has at least one even side, which is the case here, then the first player can always win. Imagine the board tiled with dominoes: 2×1 rectangles. Any tiling will do – for example, this one on the complete 8×8 board.

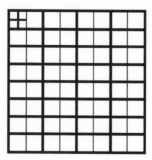

Domino grid for winning strategy on a complete board.

Whatever the second player does, other than losing by hitting the edge, the first player can always find a move for which the snake ends in the middle of a domino. That can't be a losing move, since it hasn't hit an edge, and eventually the second player runs out of options.

If both sides of the board are odd, the second player wins with a similar strategy, using a tiling that omits the initial square with + on it.

Cutting out the bottom right corner square interferes with these domino strategies. The first player can't cover the modified board with dominos, since it has an odd number of squares. The second player can't cover all but the starting square with dominos, either, but that's less obvious since those squares are even in number. But if you imagine the usual chessboard pattern of alternating black and white squares, there are 30 of one colour and 32 of the other, again omitting the starting square marked with a +. However, any domino covers one square of each colour, so a tiling would have to cover 31 of each.

There must still be a winning strategy for one or other player, because this is a finite game and can't be drawn. But it's no longer clear what that strategy might be, or who should win.

● ●

Powerful Crossnumber

^{1}5		^{2}7	7	^{3}7	^{4}6
^{5}1	2	8		^{6}2	7
2		^{7}4	0	9	6

Across

2 $7776 = 6^5$
5 $128 = 2^7$
6 $27 = 3^3$
7 $4096 = 2^{12}$

Down

1 $512 = 2^9$
2 $784 = 28^2$
3 $729 = 3^6$
4 $676 = 26^2$

• •

Magic Handkerchiefs

If you've followed the instructions correctly, the two handkerchiefs will miraculously separate.

If not, try again and be more careful.

The mathematical aspect is topological: when you convert the handkerchiefs to closed loops by clasping their ends, the loops are not linked. They just look as though they are.

• •

Digital Century Revisited

She should write down:

$$1 + 2 + 3 + 4 + 5 + 6 + 7 + 8 \times 9$$

to avoid losing her money.

• •

A Century in Fractions

The solution Dudeney asked for is $3\frac{69258}{714}$.

The others, including the example I gave when posing the puzzle, are:

$$96\frac{2,148}{537}, \quad 96\frac{1,752}{438}, \quad 96\frac{1,428}{357}, \quad 94\frac{1,578}{263}, \quad 91\frac{7,524}{836}$$

$$91\frac{5,823}{647}, \quad 91\frac{5,742}{638}, \quad 82\frac{3,546}{197}, \quad 81\frac{7,524}{396}, \quad 81\frac{5,643}{297}$$

• •

Proof That 2 + 2 = 4

This proof isn't a joke – it's how you do it in courses on the foundations of mathematics. It looks like the hard part is to prove the associative law, which of course I assumed. Actually the hard part is defining numbers and addition. That's why Russell and Whitehead needed 379 pages to prove the simpler theorem $1 + 1 = 2$ in *Principia Mathematica*. After that, $2 + 2 = 4$ is a doddle.

• •

Slicing the Doughnut

You can get nine pieces. Here are two possible ways.

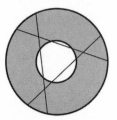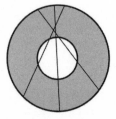

Two ways to get nine pieces with three cuts.

• •

Tippe Top Twister

When the tippe top turns over, it still spins clockwise, looking down from above.

If you imagine the tippe top spinning in space, unsupported, and then turn it upside down, it would be spinning anticlockwise. But that's not what the top does. As it starts to tip over, the end of the stalk hits the ground, and itself starts spinning. This changes the behaviour when the top finally ends up standing on its stalk.

The physical point here is angular momentum (see page 30), a quantity associated with a rotating body, roughly equal to its mass times its rate of spin round an appropriate axis. The angular momentum of a moving body is conserved – does not change – unless some force such as friction is acting.

Most of the angular momentum of the tippe top arises from the spherical part, not the stalk. Since angular momentum must be conserved – subject to a small loss through friction – the final direction of spin has to be the same as the initial direction. Friction just slows the spin down a little.

• •

Juniper Green

No other opening moves in JG-40 can force a win. There is a similar strategy for JG-100, which I'll explain in a moment, and Mathophila wins. As for JG-n, let's save that for a bit.

This game seems first to have appeared in a course on number theory given by the great mathematical physicist Eugene Wigner at Princeton in the late 1930s. More recently it was reinvented independently by Rob Porteous, a schoolteacher, to teach young children about multiplication and division. Porteous's pupils discovered that Mathophila can always win JG-100 if (and only if) she starts with 58 or 62.

The analysis depends on prime numbers, which divide into several kinds: big primes greater than 100/2 (53, 59, 61, 67, 71, 73, 79, 83, 89, 97), medium large primes between 100/2 and 100/3 (37, 41, 43, 47), medium primes between 100/3 and 100/4 (29, 31), small primes less than 100/4 but not *too* small (17, 19), and very small primes (2, 3, 5, 7, 11). The winning openings are twice the medium primes. For example, here is the analysis if Mathophila opens with 58.

Move	Mathophila	Innumeratus	Mathophila	Innumeratus
1	58		58	
2		29		2
3	87		62	
4		3		31
5	51		93	
6		17		3
7	85		51	
8		5		17
9	95		85	
10		19		5
11	57		95	
12		1		19
13	97		57	
14		LOSES		1
15			97	
16				LOSES

In his number theory course, Wigner solved the whole thing by providing a criterion for winning play in all cases. The answer for JG-n depends on whether the various powers of primes that occur in the factorisation of $n!$ are odd or even.

• •

Slade's Braid

Slade's braid trick relies on a topological curiosity: his leather strip can be deformed, in perfectly ordinary 3D space, to end up braided. So all he had to do was fiddle with it under the table until it reached the braided state. I've separated the three strips in my picture to make the method clearer.

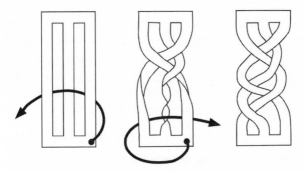

This sequence of moves introduces six extra crossings of the three strips, so by repeating it you can make really long braids.

Slade had a colourful career, and was exposed as a fraud by the Seybert Commission in 1885. See:
www.answers.com/topic/henry-slade

Avoiding the Neighbours

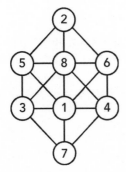

How to keep the neighbours apart.

This is the only solution, aside from rotations and reflections.

A Rolling Wheel Gathers No Speed

The point on the rim of the wheel, where it touches the road, has instantaneous velocity zero. The 'no slip' condition means that the horizontal component of velocity at this point is 0; the 'no bounce' condition means that the vertical component here is also 0.

This is interesting, because the point concerned moves along the road at 10 metres per second. But as it moves, the point on the road corresponds to different points on the wheel. And the question was about points on the wheel, not points on the road.

A more detailed analysis using calculus shows that this is the only such point. Assume the wheel starts with its centre at $(0, 1)$ and rolls along the x-axis to the right. Place a black dot on the rim, starting at the origin $(0, 0)$ at time 0.

After time t the circle has rolled $10t$ to the right, and has also turned *clockwise* through an angle $10t$. So the black dot will now be at the point

$$(10t - \sin 10t, 1 - \cos 10t)$$

Its velocity vector is the derivative of this with respect to t, which is

$$(10 - 10 \cos 10t, 10 \sin 10t)$$

This vanishes when

$$\cos 10t = 1, \qquad \sin 10t = 0$$

That is, $10t = 2n\pi$ for integer n, or $t = n\pi/5$. But at these times the dot is at positions $(2n\pi, 0)$, which are the successive points at which the dot hits the ground.

The same kind of calculation shows that any point *not* on the rim always has non-zero velocity. I'll omit the details.

• •

Point Placement Problem

It can be proved – not easily – that the process cannot continue past the 17th point.

The first proof was found by Mieczyslaw Warmus, but this wasn't published; the first published proof was given by Elwyn Berlekamp and Ron Graham in 1970. Warmus then published a simpler proof in

1976. He also proved that there are precisely 1,536 distinct patterns for placing 17 points, which form 768 mirror-image pairs.

● ●
                                        ●

Chess in Flatland

White can force a win by moving the knight.

This is the *only* opening that can force a win, but I'll omit that part of the analysis.

To see why moving the knight leads to a win, number the cells of the board 1–8 from the left. Use the symbols R = rook, N = knight, K = king, × = takes, – = moves, * = check, † = checkmate. The table shows only some of the possible sequences of moves, namely those in which White makes one move (which eventually leads to a win whatever Black does) at each step. All Black's possible replies are considered. This technique is called 'pruning the game tree', and it works provided White wins for every line of play that is included. What it omits are alternative ways for White to win, if they exist, and any White moves that could lead to a forced loss for White.

W	B	W	B	W	B	W	B	W	B	W
N–4	R×N	R×R	N–5	R×N†						
	R–5	K–2	R–6	N×R†						
			R×N	R×R	N–5	R×N†				
	N–5	N×R*	K–7	R–4	K×N	K–2	K–7	R×N†		
				N–3*	K–2	N–1	N–8†			
						N–5	N–8	K×N	R×N†	

● ●

The Infinite Lottery

You can't win. The Infinite Lottery always beats you by forcing you to remove all the balls.

This may seem rather counter-intuitive, given the way the total number of balls can increase by gigantic amounts at each step. But these amounts are finite; infinity isn't. Raymond Smullyan proved in 1979 that you always lose. His idea is to look at the biggest number in the box, and keep track of the balls that bear that number.

First, suppose that the biggest number in the box is 1. Then all balls bear the number 1. So you have to remove all the balls, one at a time – which means you lose.

Now suppose that the biggest number in the box is 2. You can't keep discarding 1's indefinitely, because they will eventually run out. So at some stage you have to discard one of the 2's and replace it with lots of 1's. Now the number of 2's has decreased. The number of 1's has gone up, but it's still finite. Again, you can't keep discarding 1's indefinitely, so eventually you have to discard another of the 2's and replace it with lots of 1's. Now the number of 2's has decreased again. Every so often you have to discard a 2, so eventually you run out of 2's altogether. But now all the balls in the box are 1's – and we've already seen that in that case you lose, however many 1's there may be.

Ah, but maybe the biggest number in the box is 3. Well ... you can't keep choosing (and discarding) 2's and 1's for ever, for the reasons we've just discussed. So eventually you have to discard a 3. Now the number of 3's drops by one, and the same argument shows that you have to discard another 3 at some point, and another, until you run out of 3's. Now the box contains only 1's and 2's, and we've just seen that in this case you lose.

Continuing in this way, it's clear that you lose if the biggest number in the box is 4, 5, 6, ..., and so on. That is, you lose no matter what the biggest number in the box is. But the number of balls in the box is finite, so there must be *some* biggest number.

Whatever it is, you lose!

Formally, this is a proof by the Principle of Mathematical Induction. This principle states that if some property of whole numbers n holds for $n = 1$, and its truth for any n implies its truth for $n + 1$, then it holds for all whole numbers. Here the property concerned is 'If the biggest number in the box is n, then you lose.'

Let's check that. If $n = 1$, then the biggest number in the box is 1, and you lose.

Now, suppose that we have proved that if the biggest number in the box is n, then you lose. Suppose that the biggest number in the box is $n + 1$. You can't keep discarding numbers n or less, because we know that if you do, you lose – that is, you run out of balls numbered n or less. So at some point you must discard one of the balls bearing the number $n + 1$, and the number of such balls drops by one. For the same reason, that number must drop again, and again ... and eventually you discard all the balls marked $n + 1$. But now the remaining balls bear numbers n or smaller, so you lose. In short, if the biggest number in the box is $n + 1$, then you lose. And that's the other step you need to complete the induction proof.

You can make the game go on for as long as you like, but it must stop after finitely many moves. However, that finite number can be as big as you wish.

. . . .

Ships That Pass ...

13 ships.

Suppose (the date doesn't matter, but this choice makes the sums simpler) that the New York ship sets sail on 10 January. It arrives on 17 January, just as the 17 January ship from London departs.

Similarly, the ship that left London on 3 January arrives in New York on 10 January, just as the ship we're talking about leaves.

So on the high seas, our ship encounters the ones that left London starting on 4 January and ending on 16 January. That's 13 ships in all.

. . . .

The Largest Number is Forty-Two

The complicated calculation is pure misdirection. The fallacy is the assumption that such a number n exists. This illustrates a key aspect of mathematical proofs: if you define something by requiring it to possess some particular property, you can't assume that 'it' *has* that property unless 'it' exists.

In this case, it doesn't.